COLIN JANSSEN AND JAN MEES

THE UNKNOWN SEA

THE IMPORTANCE OF THE OCEAN FOR PEOPLE AND PLANET

ACADEMIA PRESS

The Unknown Sea is endorsed as an Ocean Decade Activity

The Unknown Sea is officially endorsed as an Ocean Decade Activity under the United Nations Decade of Ocean Science for Sustainable Development (2021–2030). This recognition highlights the book's contribution to advancing ocean knowledge and supporting global efforts toward a more sustainable and resilient ocean. Through its insights and perspectives, The Unknown Sea aligns with the Ocean Decade's mission to transform ocean science into actionable solutions for society.

About the UN Ocean Decade

Launched in January 2021, the United Nations Decade of Ocean Science for Sustainable Development (2021-2030), the 'Ocean Decade', provides a convening framework for a wide range of stakeholders across the world to engage and collaborate outside their traditional communities to trigger nothing less than a revolution in ocean science. Throughout the Ocean Decade, partners will generate the data, information and knowledge needed for more robust science-informed policies and stronger science-policy interfaces at global, regional, national and even local levels. By collectively aligning research, investments, and initiatives around a set of common challenges, the Ocean Decade community will contribute to a well-functioning, productive, resilient, sustainable, and inspiring ocean.

2021 United Nations Decade
2030 of Ocean Science
for Sustainable Development

Colin Janssen is a full professor at Ghent University, Belgium, where he teaches and conducts research in ecotoxicology and applied marine ecology. His work explores the ecological effects of marine pollution and the intricate links between ocean health and human well-being. He has authored over 400 peer-reviewed scientific papers. He is recognized as a Highly Cited Researcher by Clarivate Web of Science and ranks among the top 1–2% of scientists worldwide in the fields of Environment and Ecology, according to leading international rankings. Janssen's research has had a profound impact on how we understand and manage anthropogenic stressors on marine ecosystems, bridging science and environmental management.

Jan Mees is the general director of the Flanders Marine Institute (VLIZ) in Ostend, Belgium, and a professor of marine biology at Ghent University. His research encompasses marine ecology, biodiversity, and data systems. He served as chair of the European Marine Board from 2014 to 2019, represents Flanders in the Belgian delegation to the Intergovernmental Oceanographic Commission (IOC) of UNESCO, and co-chairs the editorial board of the IOC's Global Ocean Science Report. With over 100 peer-reviewed publications, Mees is recognized as a leading figure in ocean science, bridging marine research, policy, and international collaboration.

CONTENTS

Foreword   9
Prologue   13

PART I ABUNDANCE   23
1. The blue planet   25
2. From Lesbos to Ostend: currents in marine science   51
3. A sea teeming with life   81
4. The ocean, mother of all life   117
5. The invisible doctor and therapist   141

PART II SWELL   165
6. The Earth's thermostat   167
7. Pollution, a multi-headed monster   199
8. Are there enough fish in the sea?   225
9. The blue acceleration   251

Epilogue   269
Acknowledgements   279
Bibliography   281

FOREWORD

It is often said that we know more about the surface of Mars than about the ocean. This is something that the United Nations aim to improve through the UN Decade of Ocean Science for Sustainable Development (Ocean Decade). As an officially endorsed product of the Ocean Decade, this book highlights all that we know and still need to know about the importance of the ocean for the comfort of humanity.

Improving our understanding of the ocean is also part of the mission of the European Marine Board (EMB), of which I have had the privilege of being the Executive Director for the past seven years. The EMB bridges the gap between marine science and policymakers, ensuring that our documents provide European policymakers with the latest marine knowledge in a format they can digest. As such, I am always editing documents on the newest aspects of marine science, which is where the privilege comes in. I get to constantly learn new things about the ocean, even though I have studied marine ecology and worked as an ecosystem modeller for more than 30 years. Since I started at the EMB, I have been educated in subjects such as the importance of the ocean for the oxygen we breathe (as is described in 'Birthplace of Life on Earth', pp. 39-41, also see European Marine Board Future Science Brief 10), and the marine geohazards that lie under the sea, and of which we know very little (see 'Chalk Cliffs and Drowned Lands', pp. 32-35, also see European Marine Board Position Paper No 26). *The Unknown Sea* also highlights the importance of the ocean for addressing the impacts of climate change (see EMB Position Paper 27), for keeping Earth habitable for humanity, but also for helping with other aspects of human health, such as mental health and well-being (EMB Policy Brief 8). It also gives an overview of the importance of the ocean for the discovery of new medical substances and highlights that the ocean has been taking up a significant part of the nutrient and plastic pollution that humans have produced.

The Unknown Sea is written by Prof. Colin Janssen from Ghent University and Prof. Jan Mees, who is the Director of the Flanders Marine Institute (VLIZ), which hosts the EMB. So I was confident that the book would be well researched and well written, but when I started reading the Dutch version of this book, it was mostly as a way to improve my Dutch. I did not really expect to learn a lot about the ocean, but I was wrong. I got a very good overview of the history of oceanography and marine science in general. There was also a nice reminder of the different explorers, which took me back to some historical ecosystem reconstruction work I did during my scientific career. It also gave me a good overview of the importance of my current hometown, Oostende, as the first place a marine station was established.

Prof. Dr. Sheila JJ Heymans FRSB
Executive Director of the European Marine Board
Professor in Ecosystem Modelling, University of the Highlands and Islands and Scottish Association for Marine Science, Scotland

PROLOGUE

> *"To stand at the edge of the sea, to sense the ebb and flow of the tides, to feel the breath of a mist moving over a great salt marsh, to watch the flight of shorebirds that have swept up and down the surf lines of the continents for untold thousands of years, to see the running of the old eels and the young shad to the sea, is to have knowledge of things that are as nearly eternal as any earthly life can be."*
> **RACHEL CARSON**

Humans are terrestrial animals. What happens in the ocean is, in large part, invisible and inaccessible to us. We are awed by tall mountains and endless seas, but many of us do not know that the tallest peaks lie submerged – we simply do not see them. Vast and mysterious, the ocean can be a source of inspiration and peace, yet it can also be intimidating and even terrifying. In the past, a sailor's 'farewell' was often a definitive goodbye.

Cartographers used to have a term for unmapped land: *terra incognita*. They also had a term for uncharted waters: *mare incognitum*, unknown sea. They would sometimes portray mythological creatures at the ends of the Earth to fill the gaps in their knowledge of the world. Today, most of the Earth's land mass has been explored. Our gaze now turns to other planets and moons, where space probes scrape extraterrestrial dust from their surfaces to expose their secrets. Our curiosity drives us ever further into the darkness of space, towards distant star systems and planets. But our knowledge of the ocean lags behind; we know more about the surface of Mars than about the depths of our ocean. These waters are Earth's final frontier of exploration and discovery.

Because of this gap in our knowledge, we underestimate the importance of the ocean for the planet and, in turn, for humankind. During lectures, lessons and interactions with the press and educators or policymakers, as well as people from various industries, coastal residents, and of course 'inlanders', we see time and time again that the sea remains a source of fascination but that our knowledge of the crucial role it plays in our lives is severely limited.

We want to change that. We wrote this book about the sea to make the latest scientific insights accessible to anyone interested in this fascinating world that begins at our shores.

As a marine scientist, you experience unforgettable moments: your first night on board a fishing or research vessel, a terrifying storm, sea and land sickness, taking samples at night, an incredible catch, your first encounter with an iconic marine animal, coming face to face with an octopus while snorkelling. And all the while, there is the wondrous light both above and below the sea's surface. We hope that you, as a reader, feel that same sense of wonder that has inspired us to – literally and figuratively – take the plunge and dive deep into the ocean.

Moreover, we want to convince you of the importance of the ocean for the continued existence of all life, including human life – not with an accusatory finger, but by sharing with you the most recent discoveries in the marine sciences and by inspiring a sense of awe when it comes to the sheer complexity of what we *do not* know.

In writing this book, we asked ourselves: what would people like to learn about the sea, including all those things they do not even know they want to learn? What does the sea do for us, and what do humans do to the sea? What has the sea meant to people and our planet in the past, what is happening today, and what do we hope to see in the future? This book is not an encyclopaedia or a reference guide; we are not striving for comprehensiveness. We want to provide the reader with what we hope is an enjoyable introduction to a world that is very precious to us, a world that is both valuable and vulnerable, and is sadly under pressure today.

TROUBLED TIMES AND URGENCY

Ever since childhood, we have loved to walk along the tideline on the beach, shifting our gaze from the timeless, ever-changing sea to the crunching of shells beneath our feet. We would collect wentletraps, cockles, and wedge clams to buy handmade tissue-paper 'beach flowers'. It was our first introduction to the world of economics with its laws of supply and demand. That tideline from our youth no longer exists. The original fauna is still there, but now you find new shell species washed up on the shore from ocean-going ships or aquaculture farms. The plastic waste cluttering up the shoreline is also new.

That does not mean all is bad news. You see far fewer oil slicks and oil-coated birds worldwide than you used to. Dumping radioactive waste in the ocean is illegal nowadays, and less unfiltered wastewater flows to the sea. And, with the growing urgency of such problems, policymakers have established fishing quotas, designated protected areas, and – in most places – prohibited whale fishing.

Still, in general, we continue to head in the wrong direction, and the negative consequences overshadow positive developments. In 2022, António Guterres, the secretary-general of the United Nations, declared an 'ocean emergency' – an emergency purely due to our pernicious influence as humans. The biggest threats are the fishing industry, pollution from land-based sources, and climate change.

What makes this problem so complex – and fascinating – is that everything in the ocean is inextricably linked. Moreover, these waters are, in turn, closely connected to the air and the land; the threat faced by one is felt by the other. This connectivity is a key theme throughout this book.

The combustion of fossil fuels on land leads to global ocean warming, seawater acidification, and ocean oxygen depletion. Melting ice caps in the north and the increasing instability of the ice caps around Antarctica – another consequence of global warming – will cause a global rise in sea levels. A strong El Niño (an intriguing weather phenomenon that we will address later) off the west coast of South America leads to heavy rainfall and mudslides in the Andes mountains and extreme drought in Australia and Indonesia. Air and sea currents carry our pollutants to the farthest corners of the ocean. We discover microplastics everywhere we look, even in the most elusive deep-sea organisms.

Still, the ocean is surprisingly robust. Its immense volume allows it to store enormous amounts of plastic, chemicals, carbon dioxide (CO_2), and excess heat. We have always been able to depend on the ocean and, as long as the planet exists, it will continue to exist. But there are limits to this capacity. The ocean may be vast, but that does not mean we should not protect it. And, because it is so enormous, we cannot ignore or neglect it.

Earlier, we referred to the ocean as the final frontier of science here on Earth. Unfortunately, in practice, it is often also the 'Wild West'. Despite the introduction of several important treaties, such as the UN Law of the Sea Convention in 1982 or the more recent UN High Seas Treaty, there is too little actual legislation, and life at sea is mostly fair game. In many so-called protected areas today, fishing is still allowed. Meanwhile, it is becoming increasingly apparent how detrimental biodiversity loss is to our ocean – and to us.

And yet, there is reason for optimism. Cutting-edge technologies and the latest scientific insights could contribute to the solution. Through the development of submersible robots, sonar systems, underwater cameras, novel sampling devices, and sensors, we should be able to access the unknown depths and the life that flourishes there. Developments in molecular biology and information technology have brought us insight into the richness of microscopic life and the functions of marine biodiversity.

All those developments have given us the tools we need to deal with the enormous challenges rising out of the water like a giant wave: the decline in biodiversity, climate change, and the pollution of our planet. We do not know for certain whether technological evolution will be enough to turn the tide. Perhaps we will have to start living differently as well. But we can be quite certain that, without this technology, we are doomed to fail.

The upside is that people finally realise that something needs to be done. The United Nations declared 2021–2030 as the Decade of Ocean Science for Sustainable Development (the Ocean Decade). At the start of that decade, a promise was made to quickly establish an operational observation system, to fill in the gaps in our knowledge and to agree on the necessary protection and recovery measures.

IN THE SAME BOAT

> *"La mer, le grand rassembleur, est le seul espoir de l'homme. Maintenant, comme jamais auparavant, l'ancienne phrase a un sens littéral : nous sommes tous dans le même bateau."*
> **JACQUES-YVES COUSTEAU**

Why are we writing this book? We have already mentioned that we share a love of the sea. It began with our natural curiosity as children and later blossomed into a professional interest. Colin is the grandchild and descendant of many generations of fishermen and seafarers and grew up in a quintessential fishing neighbourhood in Ostend, Belgium. Jan is the grandson of a Greek seafarer who settled in the Antwerp shipping quarter after falling in love with a girl who also had her roots on the Homeric island of Ithaca. He grew up in Lier, but the port of Antwerp and the mighty Scheldt River, where his parents earned their keep, formed the backdrop of his youth.

We discovered that, as eighteen-year-olds, we both shared a moment of doubt when we were trying to decide whether to attend a maritime college or get a degree in biology at university. We both decided on the latter, and it was no coincidence that we both specialised in Marine Biology at the University of Ghent, where we would later become professors. Our paths crossed again later at the Flanders Marine Institute (VLIZ) in Ostend, where Colin serves as director of the scientific board and chairman, and Jan holds the position of general manager.

Jan's doctoral thesis was about North Sea opossum shrimp, excellent swimmers that live near the sea floor and are an essential food source for bottom-dwelling fish. He later went on to study tropical fish in East Africa, and his involvement in key international marine-science networks turned him into a generalist with a helicopter view of new developments in oceanography.

Colin has devoted his career to researching the effects of pollution on aquatic ecosystems; he is internationally renowned for his award-winning research into microplastics, a study which Jan also contributed to. He is one of the 'highly cited' researchers at the University of Ghent. His current focus is on a new research field: the positive effect of the sea on human health.

We became good friends over the years. And, through our work at the University of Ghent and the Flanders Marine Institute (VLIZ), for the last forty years we have both had front-row seats to new developments in the marine sciences. We read each other's publications and act as a sounding board for each other's ideas, some crazier than others. Our most-cited co-authored paper, which we came up with during a business trip to Mombasa in Kenya, concerns the discovery of microplastics in the deep sea. This book is the result of our long friendship, our countless conversations about the sea, and our scientific partnership, which has spanned many years.

The spirit of teamwork and fellowship is what makes marine science so unique for us. It is an internationally well-organised, friendly community that is quick to pick up on new developments, including the latest technologies and open-science principles, a practice through which knowledge is shared quickly and democratically. Perhaps that has to do with the complexity and inaccessibility of the subject matter itself: the sea. As a marine scientist, you are dependent on a shared research infrastructure and also on each other. You have to be ready to give each other a helping hand during long sea voyages, sometimes in extreme conditions. We think this has made us better people. Or maybe the sea just attracts a certain type of person.

People who are in the same boat rely on each other and share responsibility. And, by extension, this applies to all of us as temporary residents on this planet as it whizzes through the immense expanse of space. As long as there is no planet B – and even if there was – we have no other option but to care for the Earth, our planet, and the largely undiscovered ocean covering its surface.

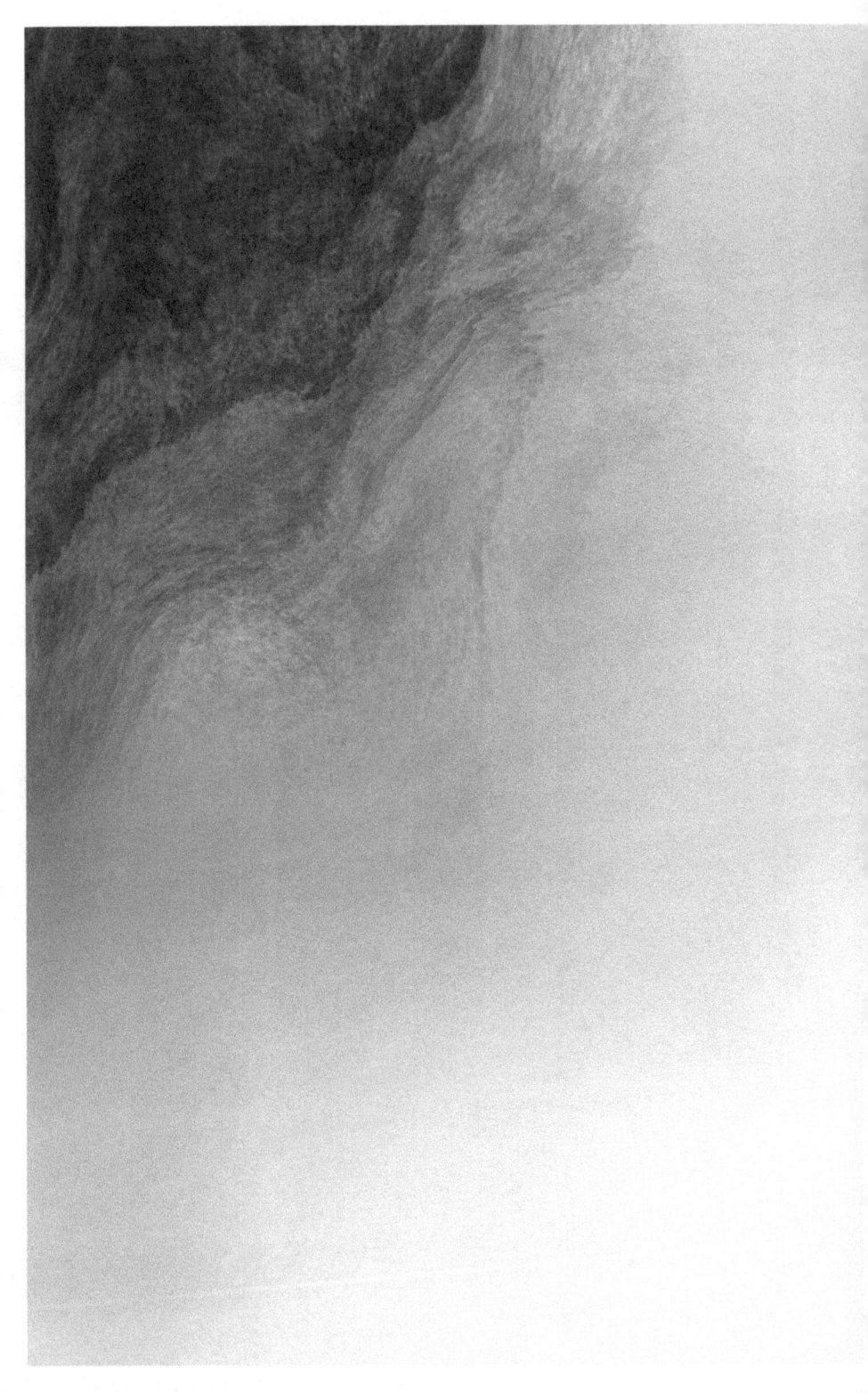

PART I
ABUNDANCE

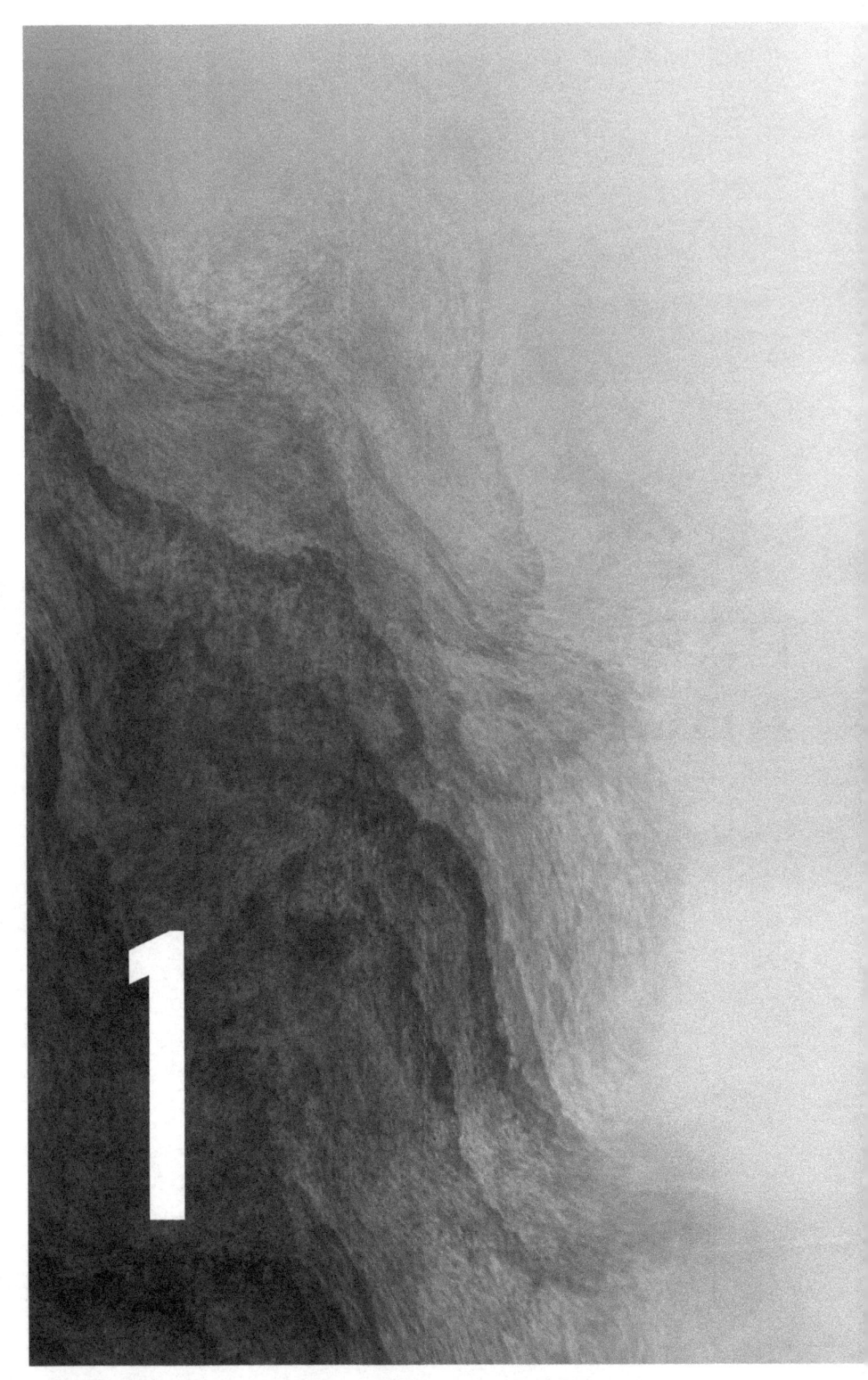

THE BLUE PLANET

> *"How inappropriate to call this planet Earth when it is clearly Ocean."*
> **ARTHUR C. CLARKE**

A pale blue dot. That is what the Earth looks like in the famous image taken in 1990 by the Voyager 1 space probe, billions of kilometres from our planet. You can barely see it; that is how insignificantly small it is. The astronomer Carl Sagan used this image as the title for his book *Pale Blue Dot*, in which he questions the hubris of humankind while urging us to cherish our place in the universe. We are not special, especially in the context of the immeasurable expanse of galaxies and twinkling stars. Everything we love or hate, all our ambitions and desires, everything that we care about takes place on that single pale blue dot.

Although the colour of our planet at such a distance is partly determined by the refraction of sunlight on our atmosphere – the reason why the sky is blue – it is no coincidence that we refer to it as the 'blue planet'. And that becomes even clearer if we imagine ourselves on a spaceship travelling through our solar system as we approach the Earth. In *The Blue Marble,* a photo of the Earth taken by the Apollo astronauts in 1972 from 'just' 29,000 kilometres away, we can clearly see how important water is to our planet: 71% of the Earth's surface is covered by ocean, a total of 361 million km².

As our imaginary spaceship draws closer and races over the ocean's surface, we would still be tempted to think that that giant body of water is nothing more than a dull, flat, two-dimensional plane. That is how many of us see the ocean. To dispel that view, we need to dive deep into this unimaginably large mass of water that has a volume of over 1.3 billion cubic kilometres and accounts for over 99.8% of all the habitable space on the planet.

Before we start to explore the obscure corners of the ocean, it is helpful if we first get a general idea of the ocean's most important characteristics. We will introduce you to seven principles drawn up by marine scientists and adopted by UNESCO to teach 'ocean literacy'. These principles will provide us with the right tools to continue our voyage on and under the water, from the coastal areas towards the deep sea, and help our eyes adjust to the dark world submerged beneath the surface.

THE OCEAN AS A CONVEYOR BELT

The first principle: Earth has one big ocean with many features

In 1992, the *Ever Laurel,* a container ship travelling from China to the United States, got caught up in a violent storm. Twelve containers tumbled into the sea. One of those containers contained unusual cargo: bath toys. No fewer than 28,000 rubber ducks and other plastic bath toys found their way to freedom, bobbed up and down in the rough seas, and started their long journey landwards. Many ducks reached the coast of Alaska. Some continued onwards to Japan, while others traversed the northern Bering Strait and were trapped in frozen polar ice before continuing their adventure and crossing over to the Atlantic Ocean. Rubber ducks were found both on the east coast of the United States and in Great Britain.

Although this was, strictly speaking, a worrisome case of pollution, scientists learned much about how plastic travelled around the world on ocean currents. For oceanographer Curtis Ebbesmeyer, these friendly floatees', as the bath toys were called, were a golden opportunity. Together with his friend Jim Ingraham, with whom he had carried out earlier research into Nike trainers that had fallen into the ocean, he called upon citizen scientists to collect as much data as possible. This gave them a better overview of the complex surface currents connecting the world's seas and ocean basins.

These globetrotting bath toys also made it very clear that the ocean is, in fact, one big whole. That is why we authors usually refer to 'the' ocean, often synonymous with 'the sea', which makes sense when you look at it from space. The five oceans we learned about at school should technically be called basins of that one ocean: the Atlantic, Pacific, Indian, Arctic, and Antarctic basins are all connected. Moreover, for practical reasons, we also distinguish between several smaller seas – their geographic indications certainly have their uses – but here too it is also impossible to define clearly demarcated boundaries.

To fully understand the concept of one ocean, we must refer to an illustration particularly close to our hearts: the

Spilhaus Projection. This image is perhaps not as famous as *Pale Blue Dot* or *The Blue Marble*, but it teaches us to look at our planet differently. That is why this is our favourite image of the Earth. This map was developed by Athelstan Spilhaus, a South African oceanographer and science communicator working in the US.

Most people are more familiar with the Mercator Projection or other typical world maps where land masses take centre stage. As the Earth is spherical and a map only has two dimensions, there will always be some distortion: no single map is completely accurate. The fascinating thing about the Spilhaus Projection is that the focal point is the ocean. It is a powerful image that succinctly illustrates the first principle of ocean literacy: there is just one ocean and everything is connected. What happens in one part of the ocean has consequences elsewhere and vice versa. The ocean, in turn, cannot be viewed separately from the rest of the planet: the atmosphere, the land, and the Earth's mantle with its plate tectonics.

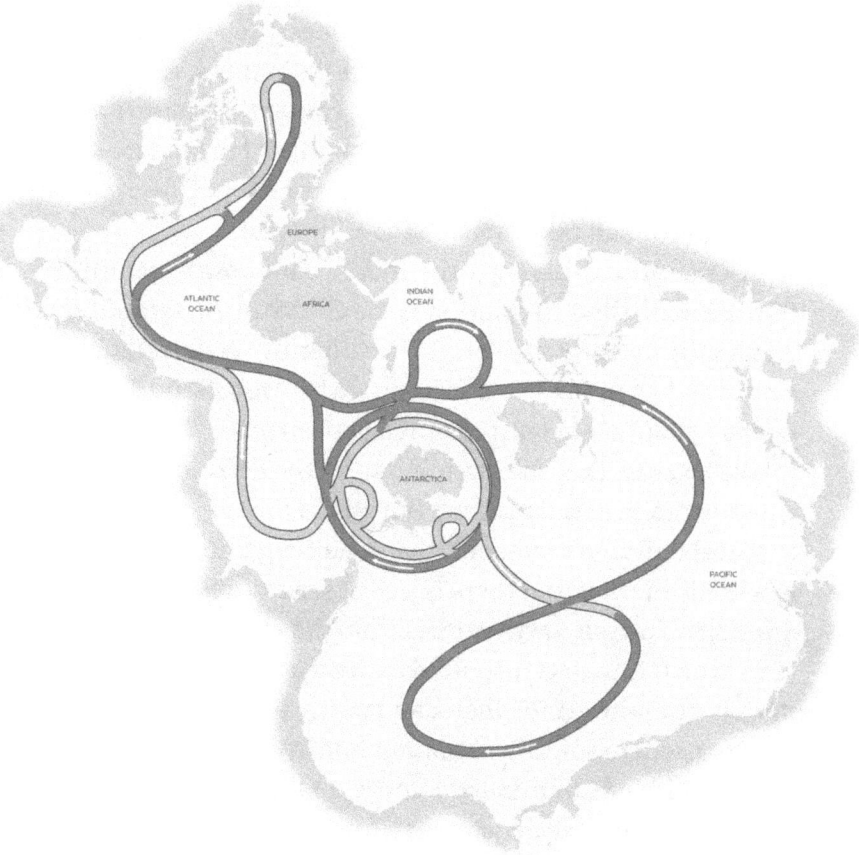

Figure 1. The Spilhaus Projection of the world, with a simplified representation of the thermohaline circulation (our addition). The dark lines indicate surface water; the lighter lines mark the deeper currents.

The map (Figure 1) also shows the thermohaline circulation. This large-scale ocean circulation transports and distributes water, heat, and dissolved substances across the planet like a giant conveyor belt, both on and deep below the surface. *Thermo* refers to temperature, while *haline* refers to salt. This global conveyor belt is driven by wind, the tides, the rotation of the Earth (the Coriolis force), temperature differences, and changes in water density (namely the salt content). Together with the shape of the ocean basins and the adjoining land masses, these factors determine the current's direction. One cycle of the conveyor belt takes about a thousand years.

A critical and perhaps the most well-known section of this thermohaline circulation is the Gulf Stream, famous for its significant influence on the weather in Western Europe. However, just in the Atlantic basin – what we call the Atlantic Ocean – there are at least 27 other vital currents.

We just saw that the 'haline' part of the thermohaline circulation refers to salt content. But why is the sea salty? This is a typical children's question that tends to get lost in an endless litany of whys, but there is a clear scientific answer to this question. The salt partly comes from the erosion of rocks on land, which is then transported to the sea via rivers or precipitation from the atmosphere, and partly from fissures in the sea floor, where volcanic eruptions and hydrothermal sources deposit all kinds of minerals (more on this later). Salt water has a higher relative density than fresh water: that is why you can read your newspaper while floating on the surface of the Dead Sea – although you still have to be careful with your paper because the water is just as wet.

The land feeds the sea and vice versa. Salt water is connected to fresh-water sources via evaporation and precipitation, while rivers provide a constant stream of nutrients, sediments, and pollutants. The idea that the sea connects everything was eloquently put into words by António Guterres during the 2022 ocean convention with the Swahili proverb *Bahari itatufikisha popote*: the ocean leads us anywhere.

Another key principle is that the seabed is not a vast, boring plain. If we could speed up time, we would notice that the ocean floor is constantly moving because of plate tectonics. The ocean holds all the planet's records: the deepest trenches, the tallest mountains, the largest plains, the longest mountain ranges, and the most volcanoes are all found underwater. The world's tallest mountain is not Mount Everest in the Himalayas, although it is the highest point at 8,849 metres above sea level. The tallest mountain rises from the sea floor: the Mauna Kea volcano in Hawaii juts out a 'mere' 4,207 metres above sea level but measures 9,966 or 10,203 metres from the bottom of the Hawaiian Trough, depending on how you measure its height. The world's largest shield volcano barely breaks the surface of the ocean. Known as the Pūhāhonu volcano, it is located about 1,100 kilometres northwest of Honolulu and is 4,500 metres tall when measured from the seabed, with a volume of about 150,000 cubic kilometres. Only a third of it juts out above sea level. In comparison, Hawaii's most famous volcano, the Mauna Loa, generally considered the largest volcano on Earth, has a volume of 'only' 83,000 cubic kilometres. With its 10,935 metres, the deepest known point on Earth, the Mariana Trench in the Challenger Deep, is much deeper than the highest point is tall.

The mid-oceanic ridges, the most famous of which is the Mid-Atlantic Ridge, form a contiguous mountain range spanning 65,000 kilometres and running through every ocean basin – much longer than the Andes mountain range, which holds the land-based record at 7,000 kilometres. Perhaps even more surprising is that you can find the tallest waterfall in the sea. On land, Venezuela's Angel Falls (Salto Ángel) plummets 807 metres to the ground. The largest falling water mass is

Victoria Falls on the border of Zambia and Zimbabwe, measuring 1,708 metres wide and 108 metres tall. But that does not include the ocean: in the Denmark Strait between Greenland and Iceland, cold, heavy water dives 3,505 metres beneath the warmer and lighter surface water, with a volume twelve times that of Victoria Falls. The longest avalanche ever recorded on Earth occurred in January 2020 in the Congo Canyon: a 1,100-kilometre sediment plume raged through the valley that leads from the Congo River to the deep sea. The avalanche reached a depth of 4,500 metres in two days.

So, the ocean is anything but a monotonous landscape; in reality, it is a spectacular phenomenon, with islands, volcanoes, canyons, valleys, waterfalls, mountain ranges, and rugged rock formations. Geological processes are not the only source of this richness; marine life is also involved in creating that landscape. Many of the rocks we know today have been and are still being formed by tiny aquatic organisms.

CHALK CLIFFS AND DROWNED LANDS

The second principle: the ocean and life in the ocean shape the features of the Earth

The white cliffs of Dover. Anyone crossing the Channel by boat from Calais in northern France to England will soon see these limestone cliffs appear in the distance. They are even visible from Cap Gris-Nez if the weather's fair enough. The English coast near Dover was once one with northern France. The two countries kissed each other like two stone lovers until they were torn apart by several catastrophic erosion events and rising sea levels. The rift between them literally ran too deep. From a geological standpoint, it was a spectacular separation: the first Brexit.

Some 450,000 years ago, the two countries were connected by a frail land bridge on a narrow chalk ridge. To the north of this bridge, a gigantic meltwater lake littered with icebergs (from ice caps in the north) lay in what is now the North Sea. A frozen, tundra-like landscape criss-crossed by rivers lay to the south. It must have been a rugged, desolate landscape with little vegetation. Over time, the ridge marking the natural dam was flooded, and for thousands of years, water flowed from giant waterfalls into what we now call the Channel, carving deep hollows at the foot of the land bridge. Scientists believe a catastrophic flood some 160,000 years ago finally cleared the Channel between Calais and Dover. It took another several thousand years before the water had risen enough to separate Great Britain from continental Europe permanently (or at least until now). We do not know the exact timing because the details are still the subject of heated scientific debate.

What we do know is this: there once existed between Great Britain and the European mainland a populated area which, like the mythical Atlantis, vanished into the sea. Scientists today refer to this region as Doggerland, after the Dogger Bank, a remnant sandbank that still lies in the Channel today. The sandbank is named after the *dogger*, a type of Dutch fishing boat that often fished for cod in the rich waters where Doggerland once lay. This long-gone land formed a broad bridge connecting England to the European mainland during the last Ice Age. Hunter-gatherers lived here some 12,000 years ago, towards the end of the last Ice Age, but rising sea levels gradually inundated the region starting about 8,500 years ago.

Around 6,200 BCE, massive submarine landslides known as the Storegga Slides occurred off the coast of Norway. They caused a tsunami that probably had a catastrophic effect on what remained of the communities in Doggerland. A higher area, which forms the shallow Dogger Bank today, was the last area to succumb to the waves and remained an island until at least 5,000 BCE.

British researchers in the early 2000s charted this area using seismic technology, the same equipment used to study earthquakes and oil fields. The results showed a hilly lowland region riddled with a complex river network, canals, and large lakes where hunter-gatherers once lived and hunted deer and other animals. We know this from archaeological finds, which included a spear tip made from antler bone, charcoal, and flint. It is an area where we will undoubtedly find many more fascinating discoveries in the years to come.

And so the sea shaped the current contours of our coastlines. The chalk cliffs of Dover and the French Opal Coast literally come from the sea. The origins of these shores date from the Cretaceous period, long before the Channel was formed. That geological period ended spectacularly 66 million years ago with the meteorite that drove most of the dinosaurs to extinction – with birds, descendants of a particular dinosaur group, being one of the few species to have survived. By the end of the Cretaceous period, a vast sea covered Great Britain and large parts of Europe. Billions of microscopic organisms lived in that giant sea, including coccolithophores, tiny single-celled algae covered with a limestone shell that sank to the sea floor when they died and formed a chalk (calcium carbonate) sediment. It is no surprise that many fossils are found in this layer. Through the same geological process that formed the Alps some 44 million years ago, the seabed rose from the depths and formed the chalk cliffs we know today.

Most of the rocks we see on land were once sediment deposits in the ocean; they were often transformed into silicate or carbonate rocks. The sea varyingly claims the land, forming inland seas, and recedes again, freeing up the land once more, which is what happened at Dover. The sea gives and takes. The constant pounding of the waves and the winds erodes the rocks, making it possible for us to soak up the sun on soft, sandy beaches in the summer.

BLOWING OFF STEAM

The third principle: the ocean is a major influence on weather and climate

2005 was a catastrophic year for residents of America's east and west coasts, especially those on the Gulf of Mexico. No fewer than four devastating hurricanes ravaged densely populated cities along the coast: Emily, Katrina, Rita, and Wilma. They were all classified as category-five hurricanes, the highest possible category on the Saffir–Simpson scale, which is used to measure tropical cyclones. The residents of New Orleans were cruelly reminded of the brutal power of nature, specifically the interaction between air, water, and land.

Sadly, such extreme weather phenomena will become more frequent as global warming continues, and they will inevitably strike close to home, no matter where you live. Think of the recent cloudbursts or 'water bombs', floods, periods of drought, and heat waves on land, in Mediterranean Europe, Africa, and Asia.

The interaction between the sea and the atmosphere drives the weather and regulates the climate in the long term. The ocean has a tempering influence on climate across the globe, acting as a buffer for the radiation and heat from the sun. How do extreme weather phenomena such as hurricanes still come about, then? The ocean and the atmosphere exchange heat, driving the water cycle and atmospheric circulation. When you heat a cooking pot filled with water, steam forms. An ocean that causes a downpour or hurricane does something similar: the sea needs to blow off steam, as it were, although it is highly unlikely that the water will ever reach boiling point and create a giant bouillabaisse. But it does sometimes reach the temperature of a warm bath: in 2023, the seawater off the coast of Florida heated to 38°C, a world record with disastrous consequences for life at sea, as we will see later.

Just as the water in the cooking pot absorbs the heat from the stove, so the ocean absorbs no less than 90% of the heat generated by the current greenhouse effect (caused by human activity). Moreover, the ocean has sequestered about a quarter of the extra CO_2 in the atmosphere through what we call 'carbon pumps'. You can see why the sea is a vital buffer to slow down and mitigate the effects of climate change.

To understand how this massive process works, we must focus on the smallest marine life on our planet, phytoplankton.

Phytoplankton comprises microscopically tiny organisms that carry out photosynthesis like plants do on land. Phytoplankton not only produces oxygen, but it is also the engine driving what we call the biological carbon pump. During this biological process, the phytoplankton at the surface absorbs the dissolved CO_2 into its cells, converts it into biological material through photosynthesis, and stores it so that it can eventually reach the deep-sea floor. When the plankton dies, it combines with other materials – such as faecal matter from marine herbivores that eat the phytoplankton – to form what is called 'marine snow' or 'ocean dandruff': carbon-rich flakes that clump together and slowly drift down to the sea floor and remain buried for long periods.

Larger animals also play a role in this process. When jellyfish die, they sink more quickly because of their greater mass. And, while this may come as a surprise, jellyfish are plankton, too. The name 'plankton' comes from the Greek word *planktos*, meaning 'drifter'. Plankton encompasses all organisms that cannot move independently against the current. Just like an autumn wind will scatter tree leaves in all directions without the leaf having any control over its direction, plankton drifts helplessly on the sea currents. The biologist Ernst Haeckel coined the Greek word *nekton* – 'to swim' – for all marine animals that *can* swim against the current, but the term is hardly ever used outside academic circles.

In addition to jellyfish, whales also play a critical role, to such an extent that some refer to the 'whale pump'. The clouds of nutrient-rich faeces released by the whales encourage the growth of phytoplankton. And when their carcasses sink to the bottom, an enormous source of carbon lands on the sea floor. According to some sources, every whale has a carbon storage capacity equivalent to thousands of trees. These are just some ways in which the biological carbon pump transfers carbon from the surface water to the deep-sea water.

The fact that we refer to it as a 'biological' carbon pump suggests that there are other ways. This pump is part of the overarching 'oceanic carbon cycle'. The other two mechanisms are the 'solubility pump' and the 'carbonate pump'. To make things more complicated – because nature is not always straightforward – these three pumps sometimes work together, and sometimes they do not. Nonetheless, all three mechanisms ensure that carbon dioxide is absorbed from the atmosphere, is distributed across the ocean, and eventually finds its way to the deep-sea floor.

The solubility pump is a relatively 'simple' process. Carbon dioxide is dissolved in seawater to form dissolved inorganic carbon, which hitches a ride on the thermohaline conveyor belt. The dissolved carbonates make the water more alkaline (basic) and act as a buffer against acidification. The solubility decreases as the water temperature increases.

The final mechanism is the carbonate pump, which is actually a part of the biological pump. This is where the coccolithophores from the chalk cliffs come in: they also belong to the phytoplankton group. Together with other organisms such as corals, starfish, oysters, and mussels, they produce calcified skeletons made from calcium carbonate. When they die, the calcium carbonate sinks to the deeper layers of the ocean, forming a cement-like layer of calcified skeletons and sediments that eventually turns into limestone.

The ocean stores CO_2 using different mechanisms. If you are still fuzzy on the details, that is fine. We will address this extensively when we talk about climate change in Chapter 6. At that point, we will also see the problem that arises as the ocean gets warmer, more acidic, and more oxygen-deprived. For now, it is enough to realise that the ocean, with its innate capacity to store heat and carbon, is an essential ally in our fight against climate change. On the other hand, if we continue to interfere with the complex interaction between the atmosphere and ocean, feedback loops can lead to dramatic and unforeseen consequences.

THE BIRTHPLACE OF LIFE ON EARTH

The fourth principle: the ocean makes the Earth habitable

Life comes from the sea. Many ancient Greeks held this opinion, following the natural philosopher Anaximander from the 6th century BCE. Although there was no scientific method at the time to support that conclusion, today there is strong evidence that they were right.

Although there is much discussion about the details and even about the main points, most scientists generally agree that life originated in the ocean. To address such a complex puzzle, we will try to figure out exactly how it happened in Chapter 4 – we need quite a bit of room to do such a monumental question justice. For now, we will focus on how the ocean has made the Earth a habitable place.

Life originated almost four billion years ago – or at least, that is the current thought on the subject; the actual date may be older. And, although the circumstances may be completely different, it is the ocean that ensures our planet remains habitable today. That is no coincidence. One of the most essential conditions for animal life is the availability of oxygen, a gas that was hardly present in the atmosphere when our planet was initially formed from space debris.

Although the Amazon rainforest is known as 'the lungs of the Earth' – a well-meant but misleading term from a scientific standpoint – the global ocean is far more critical when it comes to oxygen production. Most of the oxygen we breathe comes from the ocean, although you would be wrong to think that it was produced recently because marine life almost instantly uses up most of the oxygen as it is produced. The truth is far more spectacular: a great deal of the oxygen we breathe was created millions of years ago by phytoplankton, those microscopic algae that we became familiar with when we talked about the biological pump. One of the most important of these is a cyanobacterium called *Prochlorococcus*.

The production of oxygen, which probably started some 2.4 billion years ago, was quite revolutionary. Like other revolutions, not everyone and everything benefitted from it: oxygen was positively toxic to all anaerobic life (creatures that live without oxygen). That is why geologists, appealing to their sense of drama, sometimes refer to this as the 'oxygen crisis'. It is to oxygen that we owe our atmosphere's ozone, which protects us and all other life from hazardous ultraviolet radiation. Without ozone, there would never have been life on land. Ultraviolet radiation itself is responsible for creating ozone: it causes oxygen atoms (O_2) to split into two, and one of the split oxygen atoms binds to another O_2 molecule to form ozone (O_3). This ozone layer is an important example of how humans can severely impact the Earth: harmful CFCs (chlorofluorocarbons) from products packaged in aerosol cans were responsible for the hole in the ozone layer. Fortunately, the ozone layer quickly recovered after these harmful substances were banned.

Once oxygen appeared on Earth, countless new species flourished. The evolution to complex multicellular life would probably have been impossible without oxygen. All significant leaps in the evolutionary process came from the sea, including the colonisation of land. This led to the evolution of those evocative dinosaurs, birds, mammals, and eventually primates and humans. Over time, *Homo sapiens* became the only surviving human species. All the animals and plants living today stem from organisms that lived in the sea in ancient times.

The sea provides water, oxygen, and nutrients, all the conditions required for life on Earth. Moreover, the sea tempers the climate, making the planet habitable. We do not always fully realise it, but without that biodiversity, with its complex ecosystems and food sources, humans would never have existed.

A SEA FULL OF EXTREMES

The fifth principle: the ocean supports a great diversity of life and ecosystems

Anyone who has ever swum in the ocean and has found themselves caught unawares by a wave will recognise the feeling: the unpleasant burning feeling in the back of your throat as you swallow a mouthful of briny seawater. What you probably do not know is that this mouthful of seawater contains much more than just water and salt. It includes minerals such as magnesium, calcium, sulphur, and potassium, as well as gases such as CO_2, oxygen, and nitrogen. But you are also swallowing countless organisms: algae, fungi, viruses, thousands of bacterial species, and even tiny snails, crustaceans, and worms. Today, we know that we can categorise this varied kaleidoscope of organisms under the convenient blanket term 'plankton'.

Most marine life consists of tiny organisms (microbes), including bacteria, protists, fungi, and archaea. If you were to weigh all life on Earth, those microbes would be responsible for 90% of that weight. They form the basis of all food chains. They are the most important primary producers on Earth and are an essential part of ecosystems as they absorb and recycle food and nutrients. They keep the ocean clean.

In addition to the tiniest creatures on Earth, the ocean is also home to the largest animals that ever lived: blue whales. These enormous differences reflect the vastness of the ocean and its extreme gradients and conditions that harbour a mind-boggling diversity of life forms. Many animals only live in the sea, and there are countless life forms and ways of life that you will not find on land. Although the sea may contain fewer species, from a taxonomic standpoint, the diversity is far greater than on land: the ocean contains more fundamentally different types of species. Just to be sure we do not offend any land biologists: in terms of numbers, we are talking about the high-ranking taxa, the higher levels on the biological classification scale.

How is that possible? One crucial factor is time, as will become apparent in our chapter on the origins of life. Moreover, the ocean's capacity for life is immense, much greater and more diverse than on land. Ecosystems vary from those near the surface to the deepest, darkest depths and even beyond, to the bottom and the ocean's underworld. There is life in even the most extreme environments, in contrast to what we believed earlier. The typical ingredients that determine life are the level of dissolved oxygen, salt content, temperature, acidity, light, nutrient availability, depth and pressure, soil type, and water circulation. There are even deep-sea ecosystems that do not need sunlight and photosynthesis. Hydrothermal vents, for instance, are dependent on chemical energy and chemosynthesis instead. They provide shelter to life forms that can hardly be compared to those on land. However, our coastlines also feature a tremendous diversity in ecosystems, determined by factors such as tides, wave energy, predation, competition, and substrate type.

At the same time, the ocean is highly asymmetric. A small part is home to the richest life on Earth, with well-known iconic species, while other vast stretches seem to contain far less diversity. The philosopher Blaise Pascal once said the following about the universe: 'The eternal silence of these infinite spaces terrifies me.' This quote could, at first glance, also apply to part of the ocean, although we now know with hindsight how many hidden microbes a mouthful of seawater contains. Furthermore, we are becoming aware of how much marine life is actually hidden in the deep sea, much more than scientists ever thought possible. We will come back to this in Chapter 3.

TREASURES FROM THE SEA

The sixth principle: the ocean and humans are inextricably connected

Ever since the first humans headed towards the coast, they benefitted from the sea. We know for certain that we owe our existence to the sea. Not only because all life on Earth originated there but also because it supplies us with enough oxygen and fresh water (most rainfall comes from the sea).

For hundreds of thousands of years, our ancestors were consuming fish, a rich source of protein. In Kenya, traces of fish and sea turtles were found that were prepared with stone tools almost two million years ago. But our ambitions did not stop there. For thousands of years, we have also been travelling the ocean in wooden ships to explore new horizons, gain knowledge and – let's be honest – pillage other lands.

We are still heavily dependent on the sea today. It provides us food through fishing and aquaculture, minerals through sand and gravel extraction, and energy through windmills. Cutting-edge marine technology is widely implemented in research into energy sources, such as floating solar panels and generators that convert wave and tidal energy into electricity. The power of the waves and the tides is immense, so it makes sense for us to look for ways to harness that power to produce sustainable energy.

There are other, more subtle ways in which the ocean can provide energy. The difference in temperature between warm surface water and cold, deeper water could be converted into electrical energy using heat exchangers. We could even draw energy from the difference between salt and fresh water through osmosis. Fresh water contains less salt than seawater and will move across a membrane to an area with more salt. The resulting movement could be used to drive a turbine to generate electricity.

Countless gas and oil reserves exist beneath the seabed. But we would be well advised not to make this public knowledge because we desperately need to move away from fossil fuels if we want to do something about climate change. A guaranteed safe and sustainable extraction process does not exist, as catastrophic oil spills have proven time and time again. As long as companies continue to hunt for oil and gas at the bottom of the sea, severe accidents with catastrophic environmental impacts will continue to occur. It is simply too risky a business. We believe that the same applies to planned deep-sea mining and the search for manganese nodules and their precious metals, such as copper, cobalt, nickel, and manganese. These metals are abundant on land, but some companies and governments would like to harvest them from the seabed under the guise of a 'sustainable energy transition'. We will come back to this in Chapter 9 on the 'blue acceleration'.

Other treasures are thankfully more easily accessible. We have discovered some of them only recently, or we are still in the process of uncovering their secrets. Coral reefs are ecosystems where residents produce numerous chemical substances to repel enemies or reward allies. They are the pharmacies of the sea, as we will see in Chapter 5. In other parts of the ocean, we also see organisms with interesting chemical substances that may inspire medical practices or the design and production of biomaterials. In short, this type of research has a promising future – yet another reason to treasure our coral reefs in addition to all of the other natural habitats that host marine biodiversity.

Aside from raw materials and natural treasures, the sea provides us with economic benefits, creating jobs and maintaining national economies, especially in countries that thrive on coastal tourism. In some countries, coastal tourism is responsible for 60% or more of their gross national product. The blue economy is on the rise globally: if the ocean were a country or a state, it would be ranked as the world's seventh-largest economy. And those of us who remember the *Ever Given*, the container ship that blocked the Suez Canal in 2021, will realise how important the sea continues to be for international trade.

But our nonchalant treatment of the ocean threatens the benefits it offers for humans while exacerbating the destructive power it still holds. According to the United Nations, 40% of the world's population lives within 100 kilometres of the sea. Natural disasters such as tsunamis, hurricanes, and storm surges are becoming more frequent as a result of human interference. Individual and collective action is needed for the sustainable management of the ocean for all of us and for everyone who comes after us, even if it is out of self-interest.

DIVING TO NEW DEPTHS

The seventh principle: the ocean is largely unexplored

As we mentioned in the prologue, the ocean is the least-researched place on Earth. Although sonar and video recordings have helped study the topography of the seabed, only 25% of the sea floor has been mapped. We even have better charts of the moon than we do of the bottom of the sea.

When it comes to marine life, that percentage is much lower. A striking example is the study of life at the bottom of the deep sea. Scientists usually gather information using grabs or corers that take and carry small volumes of sediment to the surface. Researchers estimate that if we combine all the samples that have been taken from the sea floor since the *Challenger* expedition, we would not have enough to cover five football fields. The deep sea is *billions* of football fields in size. The percentage of life on the seabed that has been studied is estimated to be 0.0017%. And that is just the top 10 centimetres of the bottom. It does not include any animals that are too large or too fast to be captured by grabs, or life that may be buried deeper in the ocean floor.

We believe that, as of this moment, less than 20% of the larger marine organisms have been spotted and described. And we know even less about microbial life. We have barely caught a glimpse of their distribution in space and time, or understood the role they play in various ecosystems. If we want to gain further insight into the deep sea, we should also consider the time scale: some metabolic processes take seconds, tides take hours, biorhythms take days and nights, seasonal patterns take months, life cycles take years, and climate patterns take decades and even centuries. If we look at the ocean from the perspective of a human-centric time scale, we will miss out on a lot.

The good news is that future generations will have much to discover and that the instruments used to make these discoveries keep getting better. Oceanography has had the wind in its sails since the 19th century and has gained the ability to discover new horizons. Thanks to robots, cameras, sensors, sonar systems, remotely operated vehicles, and other technological tools, we can bring to light worlds that were once shrouded in mystery. These tools have made the unfathomable depths accessible to us. And, with advances in molecular biology, microbiology, and information technology, we are gaining insight into the richness of the sea and the functions of its biodiversity. We are slowly charting our seas, like cartographers once charted our lands, one piece at a time. Mathematical models help us make predictions based on natural phenomena and human activity. We are also beginning to understand the crucial role the ocean plays for the Earth and the climate issues we have caused.

Scientists do not have to be driven solely by altruistic interests. There is much at stake. Gaining insight into the ocean is essential for us to develop a better understanding of its interactions with the atmosphere and the land, and to help us deal with global problems.

To get a better idea of what we know now and how far marine science has come, and to understand why it took us so long to reach cruising speed, we need to travel back in time a few thousand years. Exceptional thinkers and scientists, driven by a sense of wonder and a desire to learn, repeatedly shed new light on the dark mysteries of the ocean. Without their pioneering work, marine science would have looked very different today.

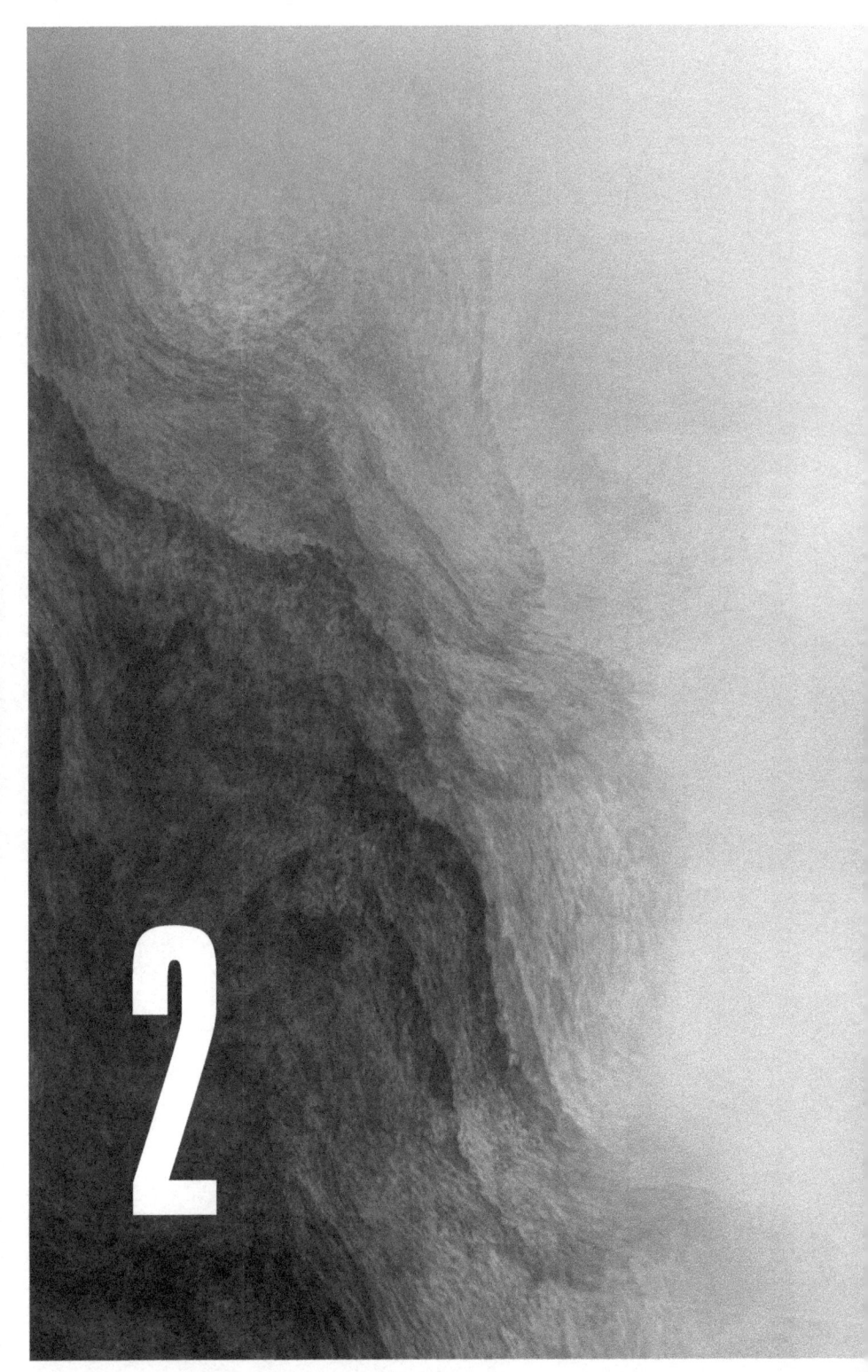

FROM LESBOS TO OSTEND: CURRENTS IN MARINE SCIENCE

> "What did the deep sea say? Tell me, what did the deep sea say? It moaned and it groaned. And it splashed and it foamed. And it rolled on its weary way."
> **WOODY GUTHRIE ('WHAT DID THE DEEP SEA SAY')**

Although oceanography as a science is, strictly speaking, relatively recent, its roots date back thousands of years. As soon as people took to the seas to fish, trade, and explore the world, we acquired the knowledge we needed about the sea and its riches and dangers, knowledge that formed the foundations on which we built the science we know today. This was often a gradual process, as insights were passed down from generation to generation. Still, there were a few individuals who pushed those boundaries further. And there was one soul who was so far ahead of his time that it took science centuries to catch up to him. To meet this seminal pioneer of marine science, we must travel to the rugged coastlines of Greece. We can think of worse places to start our journey.

ARISTOTLE'S LAGOON

A refreshing sea breeze provides welcome relief from the stifling summer heat as waterbirds fish for food in the shallows. Fishing boats bump against the jetty next to beaches where summer tourists doze in loungers under the searing sun. We are in the Gulf of Kalloni on the Greek island of Lesbos. The water's colour shifts from azure blue to purple as the sunlight reflects off the surface. It is an excellent location for birdwatchers: with a bit of luck, you can see flamingos pottering about in the nutrient-rich waters.

Over 2,300 years ago, this area had a famous guest who did far more than sunbathe on the beach: the Greek philosopher Aristotle (384–322 BCE), arguably the very first scientist ever. He earned that title with his revolutionary research in this very bay, which was once home to the ancient Greek town of Pyrrha. Today, we can refer to this bay with fitting reverence as 'Aristotle's Lagoon', which the New Zealand biologist Armand Marie Leroi did in *The Lagoon*, his book about Aristotle's research on Lesbos. Leroi even claims that it was Aristotle who triggered the study of science as a whole, as illustrated by the book's subtitle: *How Aristotle Invented Science*. And, although that claim may sound rather bold, it certainly applies to marine biology. We cannot but be amazed by Aristotle's seminal contribution to the acquisition of knowledge and the scientific method long before the scientific revolution in the 16th and 17th centuries.

Aristotle was interested in what his own mentor Plato, with his theory of ideas, considered inferior in the physical world: empirical research into concrete animals. These animals included fish, squid, sea turtles, lobsters, crabs, and the myriad of fascinating creatures writhing on the sea floor. He was not above standing with both feet firmly in the mud and getting his hands – and his cloak – dirty, something he had in common with that other celebrity in the field of biology, Charles Darwin, who dedicated years of his research to earthworms and barnacles. However, in Aristotle's time, studying such humble research subjects was hardly considered the norm. His predecessors wanted to expose all of the secrets of the universe through their abstract reflections. 'We should venture on the study of every kind of animal without distaste,' Aristotle countered, 'for each and all will reveal to us something natural and something beautiful.' Ultimately, in the natural sciences, this attitude prevails over Plato's abstract reflections. Aristotle had the following telling quote to say about his mentor: 'Plato is dear to me, but dearer still is truth.'

What exactly was Aristotle doing on Lesbos in the Gulf of Kalloni? He was probably invited there by his student, Theophrastus, the father of botany. Aristotle had left Athens earlier for Asia Minor after Plato's nephew Speusippus became head of the Academy, the institution that Plato had founded. The appointment was unexpected because Aristotle was considered the Academy's best student by far. He may have harboured some resentment against the decision, although he may also have left Athens for political reasons. Aristotle had close ties with Macedonia; he would later become a tutor to Alexander the Great. But in those days, his contemporary, Demosthenes, delivered fiery, damning speeches (called philippics) against Philip II of Macedon, Alexander's father. The Macedonian rulers had many opponents in Athens at the time.

On Lesbos, Aristotle and Theophrastus divided their areas of interest: Theophrastus specialised mainly in plants and Aristotle mostly in animals, and the latter's observations were often very precise. Aristotle described how snakes mate, how cicadas emerge from the ground, how bee-eaters eat bees...

With his 'natural philosophy', he observed countless phenomena that today are divided into the fields of physics, geology, meteorology, biology, and other natural sciences. His work was rational, based on observations, and systematic. The only thing missing was experimentation. But he had an undeniably scientific approach, and his work contains much that is still considered valid today. For instance, he wrote about the water cycle and how fresh water evaporates as seawater is heated and then returns to the sea through precipitation and river runoff.

Aristotle's theory of knowledge was based on – and an attempt at improving – the general knowledge and insights of ordinary people. This is clearly visible in his marine biology studies: he was a true pioneer in consolidating existing (common) names for aquatic fauna and thinking up new names. He did so by using the catch from fishermen in the Gulf of Kalloni. His works contain no fewer than 1,400 names of sea animals from a total of 200 different taxa: he described and named sponges, jellyfish, bristle worms, molluscs, crustaceans, echinoderms, sea squirts, fish, sea turtles, and sea mammals. He mostly based their names on the morphological characteristics of the species, a practice that is still common for assigning both scientific and common names. Aristotle's legacy was colossal: he was the father of biological classification. Many of the names he adopted, such as 'Tetrapods' (four-limbed) and 'Malacostraca' (crustaceans), are still used today. He drew up recognisable descriptions of animals such as electric rays, sea urchins, cuttlefish, the octopus, the small-spotted catshark, and the argonaut (a type of octopus). He was also the first to distinguish and classify animals according to observable 'species characteristics' such as their habitat, migration behaviour, feeding techniques, reproductive methods, and anatomical details. For that reason alone, we can call him a scientist. His empirical approach was nothing less than revolutionary.

Many of Aristotle's findings continue to be relevant in marine biology and ecology today, and not only regarding taxonomy. He also claimed that life at sea was diverse (*polymorpha*) – more diverse than on land because there were more habitats. The animal habitats and ways of life he described included sessile (attached to something), in mud or sand, intertidal (between the tides), in rock cavities or empty shells (such as the hermit crab), or living close to or on the bottom as some fish do. He distinguished between animals that give birth to live young (viviparous), animals that lay eggs (oviparous), and fish that we would call ovoviviparous today. For many animals, he described their reproductive methods and timeframes, their symbiotic relationships, parasitic behaviour, and competition between and within species living in the same habitat. His feeding typology included grazers, deposit feeders, suspension feeders (plankton grazers), predators, and omnivores, another classification we still use today. He observed and described colour changes in octopuses, the behaviour of moulting lobsters, the hunting techniques of electric rays and lantern fish, the flight of flying fish, and the leaps of swordfish.

Although Aristotle had spent a large part of his life along the Aegean Sea, he had a broad knowledge of other areas around the Mediterranean. He observed how fish populations migrated, and he discovered the correlation between reproduction and feeding. One of his favourites was the cuttlefish (*Sepia*), whose internal shells can also be found on the shores of the North Sea. He dissected its reproductive organs and intestines. His analysis remained unrivalled until the 17th century, when Jan Swammerdam also discovered the cuttlefish's three hearts.

His work may also include the first description of the overfishing of scallops through the scraping of fishing tackle over the sea floor.

All in all, that is exceptionally impressive for someone who lived in the 4th century BCE. It would take centuries before this depth and breadth of knowledge would be matched. Having said that, not everything he wrote was correct. One well-known misconception, rather sexist if you look at it from today's perspective, was that women were inferior and, therefore, had fewer teeth than men. It is odd to think that the godfather of empirical science never bothered to count them himself – after all, he was married twice.

He also missed the mark several times in the marine-biology field. He placed the sea urchins and sea squirts in the same group (*Ostracoderms*) as the bivalves and snails. We now know that that was incorrect. He knew that eels migrated from the rivers to the sea, but he had no idea how they reproduced. He could not find their reproductive organs and, therefore, assumed that they spontaneously appeared out of the mud. He may have drawn an analogy with maggots living on rotting flesh: he also thought – erroneously – that they spontaneously originated there. We now know that the genitalia of eels do not develop until they reach the Sargasso Sea in the Atlantic Ocean after a long journey.

Looking back and knowing what we do today, we can also comment on Aristotle's claims about the behaviour of living creatures, although his attempt to go further than mere observation was innovative. However, Aristotle was not familiar with the most important explanatory principle in biology: evolutionary theory. He understood that animals were made to 'survive and reproduce'. To him, this purpose was essential, while contemporary biologists only use such teleological terminology as a simplification of the blind process of natural selection, which is basically without purpose. Moreover, he could not possibly have had any idea of modern genetics. Granted, with today's hindsight, he sometimes seemed to flirt with the concept of DNA in his discussions about the *potential* of every organism to actualise a *telos* (purpose). But it would be anachronistic to assume that he had any inkling of genetics, for which we can hardly blame him, having lived more than 2,000 years before Francis Crick, James Watson, and unsung hero Rosalind Franklin, the discoverers of the double-helix structure of DNA. Sometimes, this disparity in time also makes it difficult to fully understand what he meant. When he said that every animal has a 'soul' (*psyche*), he may not necessarily have been referring to the same Christian concept of a soul that we understand today.

Aristotle sometimes described observations that scientists thought were wrong or made up for centuries but eventually turned out to be correct. One notorious example is his description of a catfish species whose females lay eggs in shallow water, after which the males bar anyone else from entering the breeding site. The father fish would spend up to 50 days with his children and attack anyone who dared to trespass. Many scientists believed this to be naive and anthropomorphic until the 19th-century biologist Louis Agassiz confirmed Aristotle's observation after studying the behaviour of Greek catfish.

It is hard for us to imagine the impact Aristotle must have had on Western ideology since ancient times, especially during the Middle Ages. His brilliance sometimes also had a paralytic effect. Aristotle was so far ahead of his time that there may have been what we refer to today as a 'first-mover' disadvantage: for a long time, there was no need to develop better ideas about biology. There was more focus on integrating them within a Christian context. Thomas Aquinas simply referred to him as 'the Philosopher' because there was no doubt as to who he was talking about. In the 17th century, Francis Bacon wrote that science first had to overcome Aristotle before it could progress, as was the case with Newton's laws of physics – although we can hardly blame Aristotle for that. It only highlights how long it actually took before anyone could match his findings. That was also the case with oceanography: that hiatus would last until the HMS *Challenger* expedition. Despite its shortcomings, his work remains one of the most impressive intellectual feats of all time.

FORGOTTEN PIONEERS AND PRACTICAL WISDOM

Although Aristotle proved exceptional in numerous fields and even laid the foundation of marine biology as a systematic science, his principles did not appear out of thin air. Other scholars also had surprising insights about the sea, although they lacked Aristotle's systematic approach.

One such scholar who certainly deserves to be mentioned is the unsurpassed and undeservedly forgotten Xenophanes of Colophon (approx. 570–478 BCE). He was a travelling poet, singing songs about the good life: wine, feasts, and athletic competitions. Today, he is mainly known for his more serious work, of which, unfortunately, only a few fragments have been preserved. He is known for his scepticism of the all-too-human gods in the stories of Homer. He said that if cattle or horses could sculpt like men, the horses would create gods that resembled horses and cattle gods that resembled cattle.

More relevant to our story is what he wrote about the sea, although he thought earth was a more important element. 'All things come from earth, and all things end by becoming earth,' he said. Yet, he felt that water also played a crucial role because he believed that the sea once covered the Earth and that the land rose out of the sea. Similar wild ideas about the four elements (water, earth, fire, and air) were circulating in those days, but what earns him a unique place in scientific history is the reasoning behind his belief. Xenophanes studied fossils. In fact, he found them deep in the interior, which implied that the Earth was once very different:

> "Shells are found in the midst of the land and among the mountains, that in the quarries of Syracuse [Sicily] the imprints of a fish and of seals had been found, and in Paros the imprint of an anchovy at some depth in the stone, and in Melite [Malta] shallow impressions of all sorts of sea products. These imprints were made when everything long ago was covered with mud, and then the imprint dried in the mud."

Another of Xenophanes' contributions was that he was the first to describe and try to explain St. Elmo's fire. St. Elmo's fire is a light that appears around the masts of ships as they sail through thunderstorms. He thought that the movement of small drops of water could account for that light. This was not correct, but it was a praiseworthy attempt. In fact, the phenomenon is caused by static electricity, something that would not be discovered until modern times. Although another natural philosopher from the 6th century BCE, Thales of Miletus, did refer to static electricity: he wrote that if you rub amber (*elektron* in Greek), particles in the near vicinity move. Thales is known, by the way, for his bold statement, 'all things are from water'. Even the most ardent oceanographer would not dare to claim such a thing. However, that statement does not make Thales a budding marine scientist. Classicists and philosophers still disagree on what he meant by it. The difference with Aristotle's empirical, no-nonsense approach based on observations is clear, but it was an important step in the evolution of philosophy, a step away from mythology and towards abstraction. That is why Thales is often referred to as the first philosopher.

Today, we use the term 'myth' to describe a persistent erroneous idea, but in Ancient Greece, it formed a colourful tradition with a religious, social, or explanatory function. Taken literally, the word 'myth' simply means 'story'. The ocean also has a place in Greek mythology. The Greek Titan Oceanus was born from a marriage between Uranus (Heaven) and Gaia (Earth). When Zeus came to power after an epic battle with the Titans, led by his father Cronos (son of Uranus), Poseidon appeared on the stage. He was given rulership over the Mediterranean Sea and earthquakes. A storm at sea, a flood, a tsunami... it was all the work of the god with the trident, whom the Romans would later give the Latin name Neptune.

Still, those myths are more than just entertaining tales: we also mentioned that they sometimes serve a social purpose. In his poem *Works and Days* (around 700 BCE), Hesiod gives a fine example of practical wisdom riddled with mythology. Although he admits to having hardly had any experience at sea, he gives the following recommendation (663):

> "Fifty days after the solstice, when the season of wearisome heat is come to an end, is the right time for me to go sailing. Then you will not wreck your ship, nor will the sea destroy the sailors, unless Poseidon the Earth-Shaker be set upon it, or Zeus, the king of the deathless gods, wish to slay them."

He says that you can sail in the spring – 'when a man first sees leaves on the topmost shoot of a fig tree as large as the footprint that a crow makes'. But he would not recommend it himself because 'it is fearful to die among the waves'.

Practical knowledge about seafaring, each claim more reliable than the last, existed long before anyone ever spoke of science. The Mediterranean Sea was busy in ancient times, not only with Greek ships but also with Cretan, Egyptian, and Phoenician vessels. Sometimes they ventured beyond the Mediterranean towards *Okeanos*, where the currents were so strong that some thought it was a river that encircled the (flat) Earth. The Greek historian Herodotus, who created the first world map and maritime chart, noted that the first people to sail around Africa were the Phoenicians. In 325 BCE, Pytheas of Massalia undertook a voyage to the north. He visited the North Sea shores, reaching England and the Hebrides, perhaps also Iceland and the Shetland Islands, the Faroes or Norway. He used the North Star to determine the degree of latitude, contributed to knowledge about sea navigation, and wrote about pack ice, the polar day, and the tides, and possibly even jellyfish (such as the 'marine lung'). His work titled *On the ocean* has sadly been lost.

But our practical knowledge about the sea dates much further back than ancient times. *Homo erectus* was a seafarer, and we also know that Neanderthals sailed and fished long before modern man did. Several tens of thousands of years ago, we do not know precisely when, Polynesian seafarers started to colonise the islands in the Pacific Ocean. They must have known much about the sea and navigation to cross such vast distances without compasses, sextants, and clocks. The first seafarers and explorers must have been aware of the waves, the storms, the currents, and the tides that kept their rafts and simple ships from always travelling in the same direction. They fished and dove for food, and must have realised that seawater was salty and not potable. Knowledge of the sea was important and led to prestige. They also made the first 'nautical charts', so-called stick charts made from bamboo or coconut fronds that were tied together. The length of the sticks represented patterns in the swells, currents and wave crests, while shells or knots made from coconut fibres marked the islands that disrupted these swell patterns.

EXPLORERS

We will skip ahead a couple of centuries because we have a long way to go before we reach modern marine science. It took a while before marine science reached full maturity in the 19th century. Until then, other historical and technological developments were needed to get science literally on board: seafaring expeditions were usually undertaken to develop new trade routes or conquer new areas and resources rather than to search for knowledge. They were often interpreted as voyages of discovery after the fact, in a way that would fit nicely into the historical big picture. Fortunately, that new-found knowledge was a positive side effect of those expeditions – unfortunately, they came at the cost of negative effects, such as colonisation and the exploitation of people and natural resources.

When we talk about modern explorers, we initially think of Christopher Columbus and his 'discovery' of America in 1492, but he certainly was not the first. Strictly speaking, the Polynesians had been exploring the region long before he ever set sail. Columbus was not even the first person to sail to America: that honour belongs to the Vikings, although the indigenous population of America had reached the continent thousands of years earlier on foot. The Vikings, sometimes romanticised in popular culture as barbaric warriors, travelled from Scandinavia in all directions in their distinctive longships called *drakkars*. They were experienced seafarers whose explorations took them ever further afield as their ships improved. In the 9th century, they colonised Iceland. Under the leadership of Eric the Red, they also established their first settlement on Greenland, home of the Inuit. Eric's son, Leif Erikson, and his men later reached Newfoundland – the American continent –

in 1002. However, they also explored large parts of Europe and traded in the Mediterranean and Black Sea. A group of Vikings even established the East Slavic state Kievan Rus, which later developed into a highly complex and politically sensitive history between Russia, Ukraine, and Belarus.

Later explorers of note include a combination of household names and unsung heroes. Everyone has probably heard of the Venetian Marco Polo, who travelled to the Far East in 1271. But who has ever heard of the Maghrebi explorer Ibn Battuta? From 1325 onwards, over the course of almost 30 years, he travelled more than 117,000 kilometres throughout Europe, Africa, and Asia, from Tangier to China. At that time, he was probably the explorer who had travelled the farthest in all of world history. And the Chinese explorer Zheng He will probably not ring any bells for most Europeans either. Between 1405 and 1433, he went on seven expeditions to the South China Sea and the Indian Ocean. He had no fewer than 317 ships and 317,000 men at his disposal. The Chinese have a rich seafaring tradition. The Europeans adopted many of their technical innovations, such as sails that could be controlled from the deck.

The major European expeditions started later. They set the tone for the Renaissance, new ideas, fantastic discoveries, and the expansion of the European world, as well as for exploitation and colonisation. The Portuguese were particularly ambitious. Bartolomeu Dias rounded the Cape of Good Hope in 1488. Vasco da Gama was the first to sail around Africa and reach India in 1498. The Italian Amerigo Vespucci was the first European to arrive in South America in 1500. And thanks to support from Emperor Charles, in 1519, the Portuguese Ferdinand Magellan became captain of the first ship to circumnavigate the globe. Magellan himself never returned home from his travels – he was murdered in the Philippines – but some of his crew survived and eventually reached their starting point, Seville. One of them was the Flemish sailor Roeland van Brugge, known to the Spaniards as Roldán de Argote. He was part of a small group that climbed a hill in Patagonia and discovered a passage to the Pacific Ocean: the Strait of Magellan. The bell-shaped mountain in Chile called Campana de Roldán, or Roldan's Bell, is named after him.

Many more expeditions followed in the late 15th, 16th, and 17th centuries, and great strides were made in shipbuilding, navigation, and cartography. To reach their destination, seafarers required a thorough knowledge of trade winds and sea currents.

A later famous expedition shed further light on the field of marine science. In 1831, the HMS *Beagle* set sail with a then not yet world-famous passenger on board: Charles Darwin. Darwin's 1839 book about his voyage on the *Beagle* mainly concerned observations on land, but what he saw and collected there provided work for his entire career and led to his insights into speciation and evolution. Those insights were also crucial to marine biology because, prior to his work, there was no satisfactory theory that could explain the enormous biodiversity in sea life. Darwin himself was also interested in marine biology and geology. In the 1840s, he published a work on the formation of coral reefs, and in the early 1850s, he wrote about the biology of acorn barnacles. He published *On the Origin of Species* in 1859. Together with Charles Lyell's *Principles of Geology* from 1830 to 1833 (and the work of Alfred Russel Wallace, who also developed the theory of natural selection independent of Darwin), this became the 'new science'. Based on those works, scientists could now formulate hypotheses and test them with observations.

THE *CHALLENGER*: THE STARTING SHOT

Although history tends to run a somewhat chaotic course, we as humans require a certain measure of order and the demarcation of periods to make sense of the chaos. So, if we had to pick a moment in history that officially defined the beginnings of oceanography, then the journey of the HMS *Challenger* is as good a place as any.

There had been earlier, more limited European and American expeditions, such as those of the HMS *Lightning* and the HMS *Porcupine*. One important motivation for these expeditions was the laying down of trans-Atlantic telephone lines, which required more extensive knowledge of the seabed. In the process, they sometimes came across organisms at great depths, which questioned the general assumption at the time that the deep sea was a barren wasteland.

The official starting shot for oceanographic research sounded with the departure of the three-masted corvette, HMS *Challenger*. The warship, which had been converted into a research ship, left Portsmouth on 21 December 1872. The vessel was almost 70 metres long and carried 243 men on board, including a team of six scientists led by the Scotsman Charles Wyville Thomson. It was a sailing ship: the steam engine was solely used for entering and leaving harbours, for dredging operations, and for keeping the boat in position during sampling. The expedition would take four years. Sailing around the world, the ship covered a total of 127,580 kilometres. The

HMS *Challenger* returned to the English town of Spithead on 24 May 1876 after a journey lasting 1,250 days, 713 of which were spent at sea.

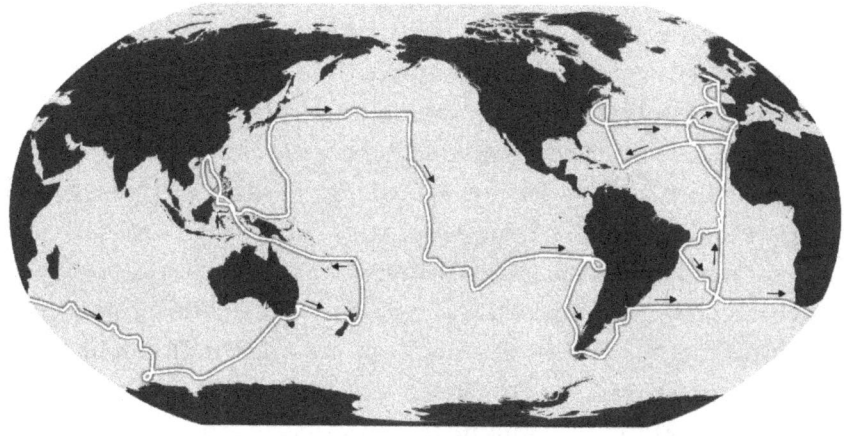

Figure 2. The scientific research expedition of the HMS *Challenger* from 1872–1876 started and ended in England.

When we look at the scale at which the scientists on board were able to conduct research, we can understand why this expedition is considered the beginning of marine science. During their journey, they described the weather and state of the sea, measured its depths, drew up temperature profiles, and sampled the water, the sea floor, and life in those waters at no fewer than 360 locations. They took almost 500 deep-sea soundings, 133 sea-floor samples via dredging, and 151 samples by trawling through the water column. The devices used to collect those samples were basic; there were no electronic devices, sensors, or sonar devices. The soundings were made with a line weighted with lead and marked with flags at 46-metre intervals. The crew could measure the sea temperature with mercury thermometers that were lowered down to varying depths.

Another reason why this expedition marks the beginning of marine science were the remarkable results. The crew later handed the samples and measurements over to the most knowledgeable specialists available at the time. Great strides were made in discovering the composition of seawater and sediments, and we owe our knowledge of sediments in large part to the Belgian scientist Alphonse François Renard.

Even more spectacular was the fact that the expedition confirmed the existence of the Mid-Atlantic Ridge, the mountain range that runs down the middle of the Atlantic Ocean like a giant spine. Some twenty years earlier, the American oceanographer Matthew Fontaine Maury suspected that the mountain range existed, based on depth soundings. The scientists discovered that the ridge runs through both the northern and southern hemispheres. Another important depth sounding was made on 23 March 1875 in what is now known as the Challenger Deep in the Mariana Trench, which lies in the Pacific Ocean between the islands of Guam and Palau. They measured a depth of 8,148 metres.

We now know that the deepest point in the Challenger Deep plunges to a depth of almost 11 kilometres. In 1960, the *Trieste*, a Swiss-Italian submersible, was the first crewed vehicle to reach the bottom of the Challenger Deep, the deepest point of the seabed on Earth. That was at a depth of 10,911 metres. The oceanographers Don Walsh and Jacques Piccard were on board. The latter was the son of Auguste Piccard, a Swiss professor at the Université Libre de Bruxelles who in the

1940s invented the term *bathyscaphe* (submersible) and created a prototype: the *FNRS-2*, named after the Belgian Fonds National de la Reserche Scientifique (National Fund for Scientific Research). Auguste Piccard was later immortalised by the Belgian comic book illustrator Hergé as Professor Cuthbert Calculus in *The Adventures of Tintin*. In 1998, a Japanese expedition also reached the bottom of the Mariana Trench with the remotely operated vehicle (ROV) *Kaiko*. The *Trieste's* crewed dive was not repeated until 2012, when Canadian film director James Cameron piloted the *Deepsea Challenger* submersible to the bottom of the Challenger Deep. Later, in 2019, the American entrepreneur Victor Vescovo reached a record depth of 10,927 metres in a deep submergence vehicle (DSV); according to the latest measurements, the bottom is at 10,935 metres.

Back to the *Challenger* expedition. The biological samples collected also contributed a great deal to the development of marine science. The scientists described no fewer than 4,700 previously undiscovered species. In those days, people did not expect to discover much life in the deep sea. The theory of evolution was still relatively new, and the prevailing theories expected to find only live fossils in the stable, silent, dark, cold, and monotonous plains without currents, waves, or seaquakes. As it turned out, that was not the case at all: the concept of a 'boring and lifeless deep sea' needed to be revised. The expedition results were published in 1895 and would eventually encompass 50 volumes for a total of 29,500 pages.

In the period after the *Challenger* expedition, scientific expeditions were carried out in different ocean basins with different ships adapted to research.

One expedition worth mentioning is the far lesser-known expedition of the German SMS *Gazelle*, which circumnavigated the globe during the same period (1874–1876). It was an answer to the *Challenger* expedition, focusing on astronomy and physical oceanography. In the spring of 1876, both the *Challenger* and the *Gazelle* berthed in the harbour of Montevideo in Uruguay, where they probably agreed to take different routes for their return voyage to Europe. Although the *Gazelle* expedition had only one scientist on board, she returned with similar observations of the same quality, yet she remained in *Challenger's* shadow. The reason was the dramatic nature of the journey: the crew suffered many health issues so there was little cause for celebration upon their return. It also took a while before the results were published.

The Dutch *Siboga* expedition in 1899–1900 to the former Netherlands Indies also proved highly successful. This expedition was led by the zoologist Max Weber, not to be confused with the German sociologist of the same name. Between 1901 and 1986, he and his colleagues published no fewer than 148 volumes on their results, which were mainly focused on zoology but also included botanical, geological, and hydrographical works.

Today, there are more than a thousand research ships worldwide, a hundred of which are more than 65 metres in length and can easily circumnavigate the globe.

OCEANOGRAPHY OF THE BLIND

In parallel to the great expeditions and the first research ships came the establishment of marine stations in Europe and North America. In addition to scientific research, there was another important reason for setting up these stations – and a motive for funding them: the decline in fish populations. Thanks to these stations, the state of fish populations could be better understood and acted upon. Small laboratories had been around since 1843, but it was mainly the German Anton Dohrn who led by example and set the course. The establishment of the marine station in Naples in 1873 did for marine biology and ecology what the HMS *Challenger* in 1872 did for 'blue' maritime research.

Still, Dohrn was not the first. Pierre-Joseph van Beneden established the very first marine station in the world in Ostend, Belgium, followed by stations in Concarneau (1859), Arcachon (1867) and Endoume (1869) in France, Kiel (1870) in Germany, and Sevastopol in Ukraine (1871). Sevastopol is situated in Crimea and is, at the time of writing, still disputed territory in the war between Ukraine and Russia.

Anton Dohrn clearly realised that proximity to the sea was an essential condition for studying marine life. He drew researchers and resources from across the globe and had a great impact on the construction of stations elsewhere. The term 'station' (the first thus named being the *Stazione Zoologica di Napoli*) was in line with the vision of creating a network of places where researchers could stop and observe local marine life. They rented out workbenches in the laboratories to foreigners and governments, and collected organisms that they sent to specialists elsewhere. The model caught on, and other countries soon followed. Other stations, still operational to this day, were built not long after: Roscoff (1872) and Wimereux (1874) in France, Den Helder (1876, now the Royal Netherlands Institute for Sea Research, or NIOZ, on the island of Texel) in the Netherlands, Edinburgh (1877), Monaco (1882),

Plymouth (1888) in England, and Helgoland (1892) in Germany, to name but a few. We saw the same movement in the United States, with its iconic Woods Hole facility where biological research started around 1871, the launching of the research ship the *Albatross* in 1933, and the biological laboratory built in 1885, inspired by the marine station at Naples.

The stations were often connected to universities located further inland. But just as often, they became places where large infrastructural marine research elements, such as research ships, were based, a role that they continue to fulfil to this day. There are currently 784 stations worldwide: 179 in Asia, 172 in Europe, 163 in North America, 81 in South America, 62 in Africa, 41 in Oceania, and 86 in Antarctica.

This pioneering period in oceanography, which lasted roughly until the technological revolution during and after the Second World War, is also referred to as the 'oceanography of the blind' because the scientists did not have the technology to visualise what lay beneath the surface. They were literally groping in the dark.

The research vessels, with their sampling and measuring devices, and the marine stations with access to the shallow coastal areas, laboratories with seawater pipes, and storage facilities, are still the workhorses of oceanography today.

The second wave in modern oceanography started during the Second World War with the development of new technologies. Just like earlier expeditions had more worldly goals with scientific knowledge as a side effect, so military developments generated new insights in other areas. Knowledge of the sea was necessary for combatting enemy submarines. Military technology, which made the hitherto inaccessible accessible and the invisible visible, found applications in science after the war, with technologies such as sonar, submersible vehicles, observation systems, and parallel innovations in information technology and molecular techniques for studying biodiversity. These innovations have dramatically changed our knowledge – and assessment – of the ocean. We know more and understand better; the technology literally sheds light on the darkness.

One milestone in the second wave of modern oceanography is the discovery in 1977 of hydrothermal vents, the so-called black smokers (although white ones also exist). Above all, the fact that life has flourished in these most inhospitable of places speaks to the imagination.

MODERN MARINE SCIENCE: A TIMELINE

The timeline below indicates several milestones in oceanographic history, with the people and events mentioned in this chapter, as well as many more milestones that we have not mentioned due to lack of space. This overview is incomplete and a matter of personal choice, but we think it is important to get a feel for the long journey marine science has taken since the first major marine expeditions and the incredibly rapid pace of evolution in the field since the Second World War.

- 1519 ~ On 20 September, the Portuguese explorer Ferdinand Magellan leaves the harbour of Sanlúcar de Barrameda for what is to become the first sailing voyage around the globe.

- 1569–1590 ~ Gerardus Mercator produces his famous world map. The Mercator Projection is his most significant contribution to the marine scientific community and shipping.

- 1586–1617 ~ Simon Stevin publishes his *De Beghinselen des Waterwichts* (a book on hydrostatic principles) and his work on marine navigation *De havenvinding* (finding harbours). In his *Wisconstighe Ghedachtenissen* (1605–1609), he describes his findings on navigation. In 1608 he publishes *De spiegheling der Ebbenvloet* (reflections on ebb and flow), and in 1617 he authors *Nieuwe Maniere van Sterctebou, door Spilsluysen*, a work on the strengthening of constructions and fortifications, and the construction of sluices.

- 1698 ~ Edmond Halley embarks on probably the very first purely scientific journey, which takes him to 52°S in the Atlantic Ocean to measure variations in magnetic compasses. He also contributes to the body of knowledge on trade winds.

- 1768 ~ James Cook explores the southern seas for 12 years and is the first seafarer to use a chronometer for determining his location (longitude and latitude) at sea.

- 1770 ~ Benjamin Franklin publishes a map of the Gulf Stream. Initially ignored, the map is only recognised when it is republished in 1786.

- 1818 ~ John Ross is the first explorer to collect water and sediment samples at great depths.

- 1831 ~ Charles Darwin embarks on his five-year journey on the HMS *Beagle*. The ship's captain is Robert FitzRoy, who also made geological observations during the voyage. In 1842 Darwin publishes *The Structure and Distribution of Coral Reefs*.

- 1840 ~ Sir James Clark Ross takes the first deep-sea sounding.

- 1843 ~ Edward Forbes declares that no life can exist in the deep sea (deeper than 500 metres).

- 1843 ~ Pierre-Joseph van Beneden establishes the world's first marine station at Ostend, Belgium.

- 1849 ~ The continental shelf along the east coast of the United States is discovered.

- 1855 ~ Matthew Maury publishes his *Physical Geography of the Seas.*

- 1860 ~ Alexander Dallas Bache produces the first detailed map of the Gulf Stream.

- 1868 ~ Wyville Thomson discovers life at depths of up to four kilometres after dredging operations from the HMS *Lightning* and HMS *Porcupine.*

- 1872 ~ Construction starts on the marine station in Naples. In 1879, the Belgian government rents a workbench at the *Stazione Zoologica.*

- 1872 ~ The *Challenger* expedition, the first major oceanographic expedition in history, sets sail on 21 December from Portsmouth.

- 1891 ~ John Murray and the Belgian Alphonse Renard classify marine sediments.

- 1917 ~ The First World War accelerates the development of acoustic techniques for tracking submarines.

- 1925 ~ The German *Meteor* expedition conducts a systematic survey with an echo sounder in the southern Atlantic Ocean and establishes the continuity of the Mid-Atlantic Ridge.

- 1931 ~ The Continuous Plankton Recorder Survey programme is launched, whereby ships of opportunity are equipped with a device that is dragged behind the vessels to automatically 'capture' plankton between two layers of fine-meshed silk. It is currently the longest-running marine biological-monitoring programme in the world.

- 1934 ~ Edward Beebe descends with a bathyscaphe to a depth of more than 900 metres, marking the beginning of manned deep-sea exploration.

- 1937 ~ Athestan Spilhaus invents the bathythermograph, a device that can continuously measure temperature and that is still used to this day.

- 1940–1945 ~ The Second World War gives rise to the development of electronic navigation systems, hydrographic equipment, deep-sea camera systems, magnetometers, side-scan sonar, and the first remotely operated submersible vehicles. In 1943, Jacques Cousteau and Emile Gagnan develop the aqualung, the predecessor to modern SCUBA equipment.

- 1953 ~ Pioneering map-maker Marie Tharp identifies a v-shaped structure running continuously through the axis of the Mid-Atlantic Ridge and postulates that this is a rift valley formed by sea-floor spreading (continental drift).

- 1953 ~ The *Trieste*, a Swiss-Italian bathyscaphe, is launched. It becomes the first manned submersible to reach the bottom of Challenger Deep in the Mariana Trench in 1955.

- 1956 ~ Marie Tharp starts drawing up maps of the sea floor based on sonar recordings. She proves that the seabed is not flat and monotonous. It is the first time we can clearly see underwater mountains, valleys, and trenches, and that the Mid-Atlantic Ridge was the likely location where the African and South American continents diverged through plate tectonics.

- 1958 ~ Charles Keeling starts taking daily measurements of CO_2 concentrations in the atmosphere at the Mauna Loa observatory in Hawaii. The resulting chart – the Keeling Curve – indicated for the first time how quickly CO_2 concentrations were rising.

- 1961 ~ The Intergovernmental Oceanographic Commission of UNESCO is established.

- 1963 ~ The first multi-beam echo-sounding system is tested, a revolutionary technique for charting the ocean floor.

- 1963 ~ The manned submersible *Alvin* is launched.

- 1965 ~ Application of the first underwater remotely operated vehicle (ROV), a cable-controlled underwater recovery vehicle. The ROV became famous when it recovered a thermonuclear bomb at a depth of 853 metres off the coast of Spain in 1966.

- 1968 ~ The Deep Sea Drilling Project begins and runs until 1983. Hundreds of expeditions retrieve almost 20,000 core samples from more than 600 locations; the deepest sample is taken 1,741 metres beneath the ocean floor.

- 1977 ~ The submersible *Alvin* discovers hydrothermal vents in the Pacific Ocean (Galapagos Rift). It is the first known ecosystem that does not rely on energy from the sun.

- 1978 ~ Seasat, the first satellite designed for remote sensing of the ocean, is launched.

- 1979 ~ Volcanic chimneys called 'black smokers', underwater hydrothermal vents where hot water (approx. 360°C) is ejected at high pressure from the sea floor, are discovered.

- 1991 ~ The Global Ocean Observing System (GOOS) is launched.

- 1992 ~ TOPEX/Poseidon, the first major oceanographic research satellite, is launched to chart the ocean. Since then, more than 100 Earth-observation satellites, many of them equipped with oceanographic remote-sensing tools, have been launched by various governments; 25 of those satellites are no longer active.

- 1995 ~ The ocean floor is mapped from space with the release of radar altimeter data from the Geosat satellite.

- 1998 ~ The Argo programme is launched, an observation system consisting of a large collection of small robot probes that are carried all over the globe on ocean currents.

- 1998 ~ The *Galileo* space probe discovers evidence of an ocean on Europa, one of Jupiter's moons.

- 2000 ~ Discovery of the 'Lost City', a white smoker field in the Atlantic Ocean. White smokers are another type of hydrothermal vent that release alkaline liquids at lower temperatures.

- 2001 ~ The Jason satellite for measuring ocean-surface topography and sea levels is launched.

- 2003 ~ The term 'ocean acidification' is introduced.

- 2020 ~ Sentinel-6 measures the ocean-surface topography and provides data for operational oceanography, marine meteorology, and climate studies.

- 2021 ~ The Decade of Ocean Science for Sustainable Development, a UN initiative, begins.

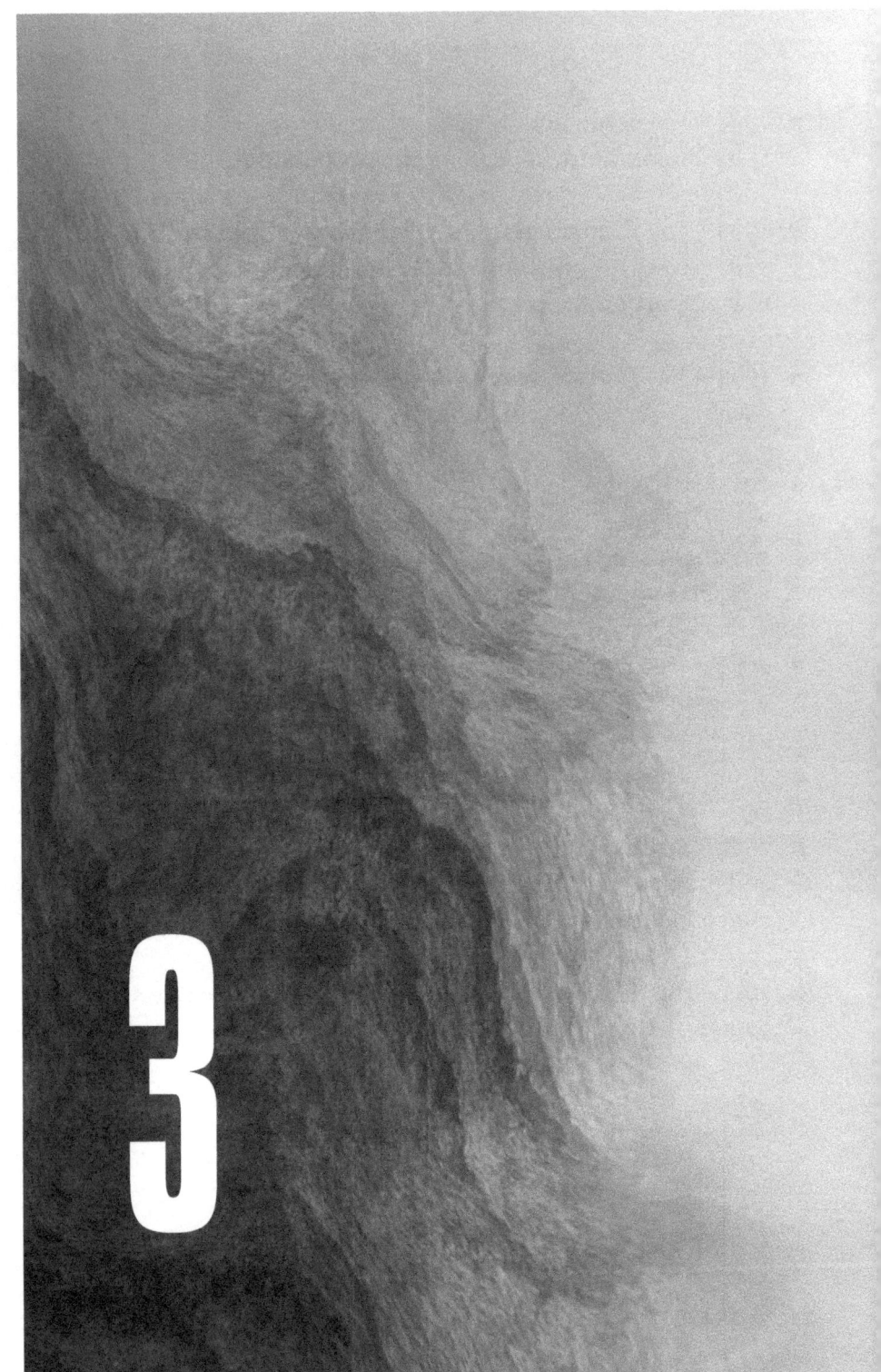

A SEA TEEMING WITH LIFE

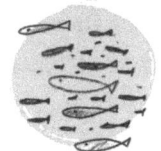

> *"I'd like to be*
> *Under the sea*
> *In an octopus's garden*
> *In the shade"*
> **THE BEATLES ('OCTOPUS'S GARDEN')**

After our journey through time and the history of oceanography, it is time to dive back into the water. We are going to continue our journey in an imaginary submarine so that we can study sea life in all its diversity in more detail. Perhaps you are already familiar with the gorgeous fish that live among the coral reefs and majestic creatures such as whales or even mythical monsters and terrifying sharks from popular films. Maybe some of these fish or shellfish end up on your plate when you enjoy a sunny day at the seaside. Serious beachcombers as well as those who enjoy a casual seaside stroll will be familiar with the shells you find along the shoreline. You might also have noticed cuttlebone: the soft, oval, chalky shell remains of the cuttlefish, a squid-like creature that also lives in the North Sea.

But there is so much more! So little is known about most marine life, usually because it is either invisible or inaccessible to us as land animals. That also explains why it has taken so long for the marine sciences to pick up speed. You can snorkel in tidal pools and shallow coastal waters, but you need special equipment to explore most other zones. Moreover, many organisms are microscopically small or are hidden deep in the seabed. Animals exist in all layers of the ocean: some lie buried in the sea floor or attach themselves to rocks, while others float or swim. Some creatures live as hermits in hidden cracks and cavities, or hide in shells. Parasites even hide inside other animals. And all that life is spread out over an unfathomably large volume of water.

If you want to study the seabed in the depths of the ocean, you need an expensive research ship to travel days or even weeks to your destination before lowering long cables, bottles, grabs, and nets to collect samples. It takes about four hours to collect such a sample, assuming that what comes up will be intact. And then all of that effort is just to get an overview of the tiniest fraction of what the seabed holds. Thankfully, the technology keeps improving, and marine science has come a long way.

Before we travel with our imaginary submarine to more exotic regions in the ocean, let's take a moment to reflect on the shores of the North Sea. You will soon discover that this sea also has a lot to offer.

THE NORTH SEA: A YOUNG BRANCH OF THE ATLANTIC OCEAN

For us Belgians, even though our short coastline runs for just 67 kilometres, it is still a beloved destination for relaxation, sunbathing, or water sports. When we stand on the beach, ignore the drab apartment buildings behind us, and focus our gaze on the blue-grey waves of the North Sea, we see a fascinating part of nature. Though, to be honest, the natural coastal habitat is far more spacious and better preserved in our neighbouring country, the Netherlands. The North Sea itself is one of the most studied seas in the world, thanks in part to Belgian pioneers such as Pierre-Joseph van Beneden and Gustave Gilson. For over 150 years, sea life in our part of the North Sea has been researched thoroughly, and we can say with some confidence that our inventory of multicellular organisms is complete. This tradition has not only produced great scientists but has also resulted in many species first being recorded in Belgium or being named after Belgians. This makes the North Sea not only an ideal field lab for our introduction into marine biodiversity but also an excellent example of the eternal struggle between land and sea.

Looking out to the North Sea from the beach, it may first appear like a moving canvas, with slashes of blue, grey, green, and white applied in thick, impasto strokes, like a William Turner painting. That interplay of colours alone is fascinating, but there is so much more. If you have never seen a nautical chart before, hopefully you will be pleasantly surprised when you see one for the first time. A nautical chart reveals what lies beneath the surface. The shallow sea is riddled with countless sandbanks with intriguing names such as Paardenmarkt, Trapegeer, or Kwintebank. Many of these names are very old.

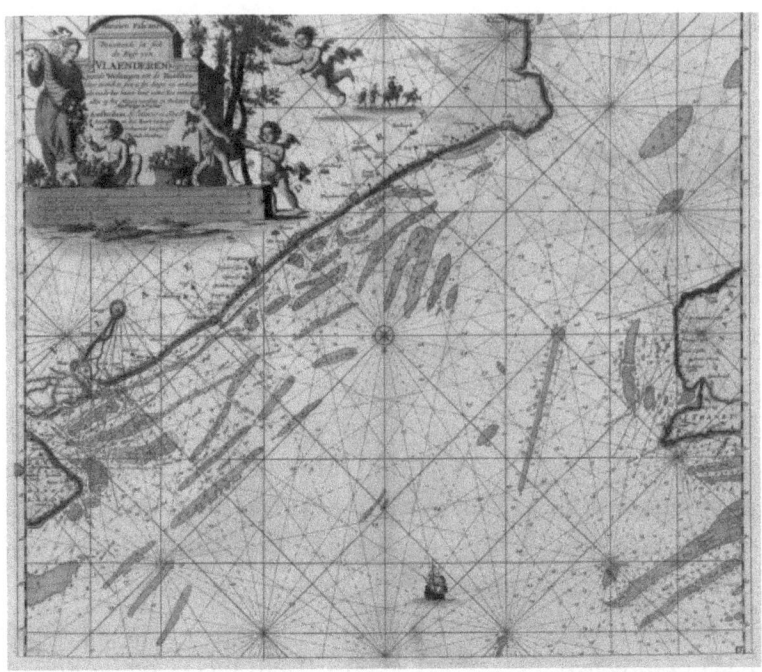

An 18th-century nautical chart with sandbanks (J. & G. Van Keulen). (French) Flanders is situated to the left, England to the right.

Trapegeer, lying just off the Belgian coast at De Panne, is a compound of old words: *geer* means 'arrowhead', while *trape* is a type of fishing tackle with a lead weight. This was an arrowhead-shaped sandbank suitable for fishing with a *trape*. The Kwintebank is probably named after an old sea tax: the *kwinte*. But it is the Paardenmarkt (literal translation: horse market) in Knokke that we want to focus on. The explanation for its name is as simple as it is surprising: this was once actually a horse market. How is that possible?

During the Middle Ages, the Paardenmarkt was probably part of the island of Wulpen, which was later swallowed up by the sea and no longer exists today. It was situated at roughly the same latitude as the Zwin nature reserve in Belgium and was inhabited in the 12th century. The island even had its own hospital and an abbey. Sadly, no written records remain but researchers assume that part of the shoal was used as a place to trade horses.

Wulpen is just one of the many places that have been swept away by the rising seas and slowly faded into obscurity. A more well-known example is Testerep, the island that once lay off the coast of what is now Ostend. The city was founded there, but later the old town centre was forced to move further inland. Sheep and cattle grazed on Testerep, and there were houses, a church, a market square, and even a town hall. When the former city of Ostend expanded and its natural defence line of dunes weakened, Testerep fell victim to storm surges. In 1394, the island was decimated by heavy flooding in the Saint Vincent storm. After barely a century, the original residents of Ostend were forced to leave the area. In the centuries that followed, the island and its ruins gradually disappeared into the sea, a prime example of nature's power. Now the island is only food for stories and archaeological research.

Paardenmarkt and Testerep illustrate how treacherous and changeable the North Sea can be – as did our story about Doggerland from Chapter 1. It is a young sea, especially along our shores. After the last ice age, some 11,000 years ago, sea levels started to rise due to the melting ice caps. The North Sea is not only young but also relatively shallow. On average, it is 94 metres deep, although that average is increased quite a bit by the Scandinavian parts of the North Sea, which are much deeper. Near the Belgian and English coast, the waters are no deeper than 46 metres and often far shallower.

The North Sea is relatively poor in species diversity due to its relative youth and shallowness. While we estimate that some 240,000 marine species have been discovered worldwide, the Belgian North Sea 'only' has 2,213 species. Of those, 227 are animals; the rest are algae. Most marine animals that live here today are opportunists that have colonised this 'new' sea fairly recently, in contrast to the older ecosystems we will encounter later.

The animals that are most striking in appearance are the fish, vertebrates just like us. Some 120 fish species live in our part of the North Sea, of which about 100 are permanent residents, while the remaining species are in the 'migratory' category. Many of these fish are well known from the fishing industry: flatfish such as plaice, flounder, turbot, and brill; herring, sprat, and mackerel from the open sea; and bottom-dwelling round fish such as the sandy dogfish and the highly endangered cod. Sadly, there are also fish species that have already become extinct due to overfishing: the angel shark or monkfish (*Squatina squatina*) and the bluefin tuna once swam here in large numbers. Other species became extinct because they needed clean fresh water to complete their life cycle: the common sturgeon, the allis shad, and two varieties of the *houting* (*Coregonus* species), members of the salmonid family. There is also a recent newcomer: the Atlantic croaker (*Micropogonias undulatus*), a croaking fish that produces its distinctive sound with its swim bladder.

About 25 sea mammal species have been spotted in the Belgian North Sea. Although most of these are temporary guests, eight of them are permanent residents: the common and grey seals; whales such as the harbour porpoise, the long-finned pilot whale, and the common minke whale; and dolphins such as the bottlenose dolphin, the common dolphin, and the white-beaked dolphin.

Molluscs are also well represented here. You may tend to associate squid more with restaurants in port towns along the Mediterranean, where they serve *calamares, calamari,* or *kalamaria* (*calamarium* means 'ink pot' in Latin), depending on the country. But we also have our own squid in the North Sea. We have cuttlefish (*Sepia officinalis*), which is a long-time resident of the North Sea whose typical calcified oval skeleton we mentioned earlier in this chapter. Our most common squid is the *Sepiola atlantica* or Atlantic bobtail, a gorgeous miniature version of the cuttlefish. Other squid species are becoming more common as well, such as the common squid (*Loligo vulgaris*). The warmer the water in the North Sea gets, the more of them we can expect. Moreover, cod and other natural enemies have become rarer due to overfishing, which means that squid have it easy here. You might be surprised to know that we are already exporting squid to Spain. On a side note, calamari is made from the common squid; the long mantle is sliced into rings. Squid are not to be confused with those other ink-producing cephalopods: octopuses. Octopuses are larger and have eight arms with suckers, while squid and cuttlefish have two longer tentacles and eight shorter arms. So, while octopuses are cephalopods, not every cephalopod is an octopus. You will rarely spot an octopus in the North Sea. In 2018, a Dutch amateur diver took pictures of an octopus off the coast of Callantsoog in North Holland.

There are also other molluscs in the North Sea, such as mussels and oysters, which are the pride of the Low Countries, along with others that are not so easy to eat: whelks, periwinkles, and limpets. Exotic species are becoming increasingly frequent among the shellfish: the Atlantic jackknife clam and false angelwing from America, the Japanese oyster, and the Manila clam are now common sights along the tidemark. These exotic species pose more and more of a threat to the local biodiversity. Their numbers have increased by 15% in less than a decade. Although improved monitoring may explain part of it, economic activity seems to be the main cause of this increase. Shipping is probably responsible for three out of every four 'successful' introductions of exotic species to the Belgian North Sea, while aquaculture is potentially responsible for four out of ten species. What is particularly striking is that almost 60% of these introductions took place after 1990, during a period when intercontinental shipping saw a rapid increase. The fact that the Flemish ports – in fact, all North Sea ports from Le Havre to Hamburg – are located along some of the world's busiest shipping routes increases the chance of new, unintentional introductions through ballast water or growth on ship hulls. Still, they are not all problematic. We estimate that only about 10–15% of all non-native plant and animal species, both on land and in the water, pose a threat to European biodiversity.

We think that some species might even have been early examples of exotic species, such as the naval shipworm (another shellfish), but we are still not sure. The naval shipworm is notorious because of the great European 'shipworm epidemic' in the first half of the 18th century, which caused enormous damage to Dutch dikes, sank ships, and caused the Slyckens locks in Ostend to collapse in 1752.

Some shellfish were almost extinct in the North Sea. The dog whelk, for instance, disappeared temporarily due to the use of ship paint containing toxic tributyltin, which caused the females to develop male sex organs and impeded their reproduction. But after this toxic, hormone-destabilising paint was banned, the population recovered. Their diet includes acorn barnacles that, remarkably enough, belong to the crustaceans. They swim freely as larvae but once they mature, they attach themselves to hard surfaces, such as mussel shells; they are the typical white crusty bumps you find on mussel shells. The hermaphroditic barnacles also proudly hold the record for having the longest penis, albeit in relative terms: during the mating season, it grows to a length of no less than seven times their body length.

For fans of larger crustaceans such as lobsters, crabs, and shrimp, no fewer than 70-odd species live in the North Sea. If we include the smaller species, such as barnacles, copepods (the bulk of zooplankton), amphipods, isopods, opossum shrimp (those shrimp from Jan's thesis earlier), and comma shrimp (named after the punctuation mark), there are countless more. Do not worry: we will not list them all here. The most famous of these are, sadly enough for the animals themselves, also the tastiest: the brown shrimp, the European lobster, and the North Sea crab. Those of you who like to eat mussels and have been brave enough to study one close-up will also be familiar with the pea crab (*Pinnotheres pisum*). It likes to live in mussel shells. And, because the mussels sometimes suffer as a result, they are considered parasites. They also sometimes live – uninvited – in the rectum of sea cucumbers. The reproductive habits of the pea crab are a bizarre spectacle: the male spends hours rubbing himself against a mussel shell harbouring a pea crab female until the mussel opens its shell and lets the male in.

Another well-known but less edible member of the crustacean family is the hermit crab. This crab lives in abandoned shells, often snail shells. As soon as the crab grows too big for its home, it moves. This sometimes leads to an elaborate game of musical chairs, whereby another hermit crab moves into the just-abandoned shell, freeing up its own shell for the next hermit crab, and so on. They also often fight for each other's precious shells.

For those of you wondering why there are so many varieties of crabs and lobsters, you are not alone: biologists have been asking themselves the same question. Interestingly enough, the body plan for crabs has evolved independently on at least five different occasions. The American scientist Stephen Jay Gould introduced a famous thought experiment in 1989 when he asked himself what the world would look like if we could turn back time and repeat the evolutionary process. Would humans have resurfaced then as well? He did not think so. But crabs probably would have. Evolution cannot seem to help but create crabs. A term has even been coined for this phenomenon: 'carcinisation', a prime example of convergent evolution (independent evolution of similar traits). We do not know exactly why nature keeps developing crab-like species, but it may have something to do with their mobility and protection from predators.

There are also countless colourful and versatile invertebrates in the North Sea, such as sponges, anemones, jellyfish and sea gooseberries, starfish, sea urchins, sea squirts, sea spiders, sipunculid worms, sea mice, sandworms, and much more. There are too many to mention, but every one of them is fascinating to observe. Note how many of these names are derived from land animals, betraying our landlubber background once more. We categorise the world according to what is familiar to us.

Although we mentioned earlier that our part of the North Sea has been extensively mapped and studied, there are many more aspects that we do not even know about yet. We know almost nothing about the fungi in the sea, the mushrooms, moulds, and yeasts, for instance. And there remains a whole world to be discovered when it comes to the smallest organisms of them all: we have yet to discover how many types of diatoms and green algae are out there. We have already discussed the critical role of phytoplankton in the biological carbon pump, but there are also less pleasant algae, some of which we are already familiar with.

When you walk along the shoreline on the beach, you will sometimes see clumps of foam, as if the waves were whipping up a meringue with egg whites. This is not such a bad analogy when you think about it because the process is very similar when you look at it from a chemical standpoint. The culprit is the alga *Phaeocystis globosa*, or more specifically, the proteins in the dead alga. This 'sea foam' is sometimes brown or green, and it often causes problems for surfers or swimmers. In 2020, five experienced surfers in Scheveningen in the Netherlands died when they were caught by surprise in a layer of sea foam metres high. Still, the algae have their uses because they serve as food for zooplankton and thus are the foundation for the entire food chain. Yet too much foam – turning the coastline into one giant cappuccino – can also be an indicator of too many nutrients, such as nitrogen, in the water.

A more pleasant phenomenon for beachgoers is sea sparkle (*Noctiluca scintillans*, which roughly means 'shining light at night'). When the weather is warm and calm, bioluminescent sea sparkle lights up the sea. In terms of beauty, this nocturnal spectacle easily matches the traditional New Year's fireworks show on the beach. The light that so entrances us is actually a pretty side effect: its true purpose is to deter predators.

The North Sea also contains bacteria that generate electricity, the long, filamentous cable bacteria that belong to the Desulfobulbaceae family. They are live batteries that can be found worldwide and which were recently also discovered in Ostend and the Dutch Delta region. They communicate with each other through electric currents, which are converted into chemical signals. One end of these bacteria is buried in the sea floor, where it collects electrons from energy-rich sulphur compounds. Unfortunately, they cannot provide the solution to the energy crisis, but who knows, they may one day inspire other practical applications.

As much as we would like to spend more time in the North Sea, it is time to continue our journey to other parts of the ocean that often contain even more diverse life. We can set course along two axes: a horizontal axis following the points of the compass or a vertical axis down into the depths. Whichever way we go, we will find an incredibly diverse range of habitats. As we already mentioned, the sea may look boring on the surface, but appearances can be deceiving. If we want to understand its complexity, we must adopt a three-dimensional approach.

There is a surprisingly broad variety of habitats. Just think of what we already know and can access along the coastline. In the open sea, there are visible surface-based communities (what we call *neuston*, organisms that live at the surface), such as Sargassum seaweed and the specialised crustaceans, sea snails, and jellyfish that live among its fronds, as well as the megafauna that occasionally surfaces such as sharks, whales, or dolphins. But there is much more: the latest Global Ecosystem Typology issued by the International Union for the Conservation of Nature (IUCN) names four marine habitats or biomes. They are: the marine- or continental-shelf biome – in other words, the coastal areas in the broadest sense of the word; the pelagic-ocean-waters biome, or open-sea biome; the deep-sea-floors biome; and – surprisingly – the anthropogenic marine biome (which includes structures like artificial islands or offshore windmills). We will explore these zones later. Our introduction will have to be brief because there are simply too many fascinating animals, plants, and even microbes in the sea. It is as if you only have a couple of days to explore a foreign country: your visit would usually be limited to a few highlights. Hopefully, these highlights will inspire you to return again later.

We can be brief about the anthropogenic zone because it was formed very recently. Humans are also a very recent addition when compared to marine life in general, although our impact has been significant. The anthropogenic zone is comprised of artificial structures that have, intentionally or unintentionally, entered the sea and attracted marine life. Examples include offshore wind farms, shipwrecks, submerged rubbish tips, offshore oil and gas rigs, artificial reefs, and aquaculture infrastructure. There are countless shipwrecks: more than 300 known shipwrecks rest off the Belgian coast alone. This human impact is a controversial topic, especially when it comes to wind farms. In Chapter 9, we will discover that these create some disadvantages for marine life but also provide unexpected benefits that offset the downsides. Groynes and breakwaters are pleasant environments for different species of seaweed, crabs, and seabirds.

CAPRICIOUS COASTLINES

While we are familiar with the shores of our own North Sea, they represent just a fraction of the diversity in coastal habits worldwide. The shallow areas along the coast are home to the sandy or rocky areas known as intertidal zones, which literally means 'between the tides'. This is the area that falls between low and high tide, and that we are familiar with from our beaches. We also have estuaries, where freshwater from the rivers meets the briny seawater. Typical habitats include mud flats (which are unsubmerged during low tide), salt marshes (which are only submerged at spring tide), and the brackish water zone. The area where sea, land, and rivers meet gives rise to fertile surroundings.

Although these are the places that we humans are most familiar with, they are relatively extreme environments: cutting winds, constantly changing water levels, fluctuating salt-content levels, searing sun, and impressive temperature differences make it hard to adapt to these fickle coastal conditions. And yet, they are brimming with life that takes advantage of the countless nutritional sources found in this habitat.

That is also the case in shallow seagrass meadows, the only marine ecosystem that features flowering plants. They are the nurseries and shelters for many animals, from the tiniest invertebrates and fish to sea turtles and large sea cows, mammals that like to graze there. Seagrass meadows are somewhat similar to kelp forests, which are also primarily found just off the coast. Kelp forests are composed of large brown algae that can grow up to 50 metres long at an astonishing rate of 30 centimetres per day. These are incredibly nutrient-rich areas that often arise where cold nutrient-rich water wells up to the surface. Not only fish but also seabirds and sea mammals love these kelp forests. They are sometimes referred to as the rainforests of the sea because they store a great deal of CO_2 through photosynthesis.

Strictly speaking, the coast belongs to the 'continental shelf' or 'marine shelf' biome. That is, the part of a continent that lies underwater. Our shallow North Sea is located on a part of the continental shelf that has not always been submerged – remember the story of the now submerged Doggerland. And, along the shores of the Atlantic Ocean, the land does not just end where the water begins. The continental shelf runs from the coastline to the bottom of the deep sea. It contains ecosystems that are, on the one hand, biological in origin, such as seagrass meadows, kelp forests, oyster banks, and coral reefs and, on the other hand, based on mineral structures, such as rocky reefs and sandy or muddy flats. The crucial factors for determining the existence of marine life are primarily the availability of light and nutrients. Light is dependent on the clarity of the water and the depth, which in turn determines whether photosynthesis by phytoplankton or larger seaweeds is possible. Many other animals are dependent on that plankton and, in many cases, they form the basis for the food chain. How effectively the phytoplankton can photosynthesise is also dependent on the currents that carry the nutrients from the depths to the surface and the availability of nutrients from terrestrial sources, especially from rivers. Coastal areas are often shallow, which means that the kinetic energy from waves (and shearing sea ice in the polar regions) plays an essential role in the distribution of species within an ecosystem. The bottom composition determines whether sessile (immobile) organisms such as coral polyps, tubeworms, oysters, and mussels can form reefs that mitigate the power of the waves. These reef builders are ingenious ecosystem engineers who create and maintain their own habitat. Other factors that decide to what degree animals and plants can create their habitats are: temperature – determined by the depth and the degree of latitude – and salinity (the salt content) – dependent on the distance from the land and its river systems. Predators also impact the habitat, as do currents, particularly for the dispersion of larvae and nutrients.

The most famous reef builders are undoubtedly the coral polyps, which work symbiotically with zooxanthellae, a type of alga, to set up a colourful ecosystem that benefits countless animals. We will focus more on this ecosystem and the unexpected benefits for human life in a later chapter about the apothecary of the sea.

THE GIANTS OF THE OPEN OCEAN

We move away from the capricious coastlines and set sail for the ocean's pelagic zone (*pelagos* is Greek for 'sea' or 'ocean'). It is by far the planet's largest biome and encompasses the open ocean's water column across all degrees of latitude.

The biodiversity is greatest in the water layers closest to the surface. Depth determines the availability of nutrients, organic carbon, and – above all – light. The more light there is, the more algae (mainly phytoplankton, but also larger algae) can photosynthesise. They form the foundation of a giant food chain. Moreover, most marine mammals, just like us, depend on light for movement and for being able to see their prey. For practical reasons, photosynthesis is limited to the top layer of the open sea, what we call the 'epipelagic' or photic zone, that still receives sunlight. This zone is only about 200 metres deep, but with the presence of phytoplankton, it is responsible for half of all carbon storage in the ocean. The zones beneath this layer are the twilight or mesopelagic zone (200–1,000 metres deep), the midnight or bathypelagic zone (1,000–3,000 metres), and the abyssal or abyssopelagic zone (3,000–6,000 metres). Everything deeper than that we call the trenches or the hadalpelagic zone after Hades, the Greek god of the underworld. Because sunlight barely reaches these deeper layers, we come across entirely different ecosystems that are mostly unknown and unexplored. We will shed some light on the mysteries of the dark, deep sea later, but let's dwell on the surface first.

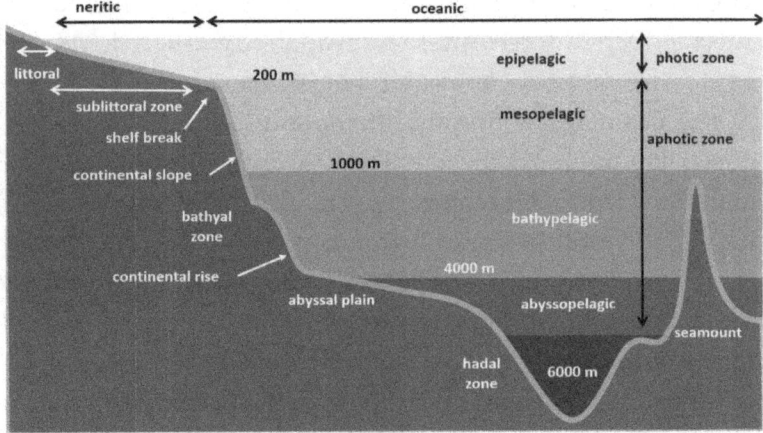

Figure 3. The different zones in the water column and the bottom of the ocean: each zone has a distinctive physical environment with characteristic processes, species, and ecosystems. Own figure adapted from Coastal Wiki Ocean Zones.

In the epipelagic (photic) zone, phytoplankton is an abundant source of food for zooplankton. The smaller fish that eat the zooplankton are, in turn, eaten by larger fish that continue up the food chain, hence the proverb 'big fish eat little fish'. The most successful big fish include the tuna varieties, which provide us with a delicious meal but are notorious and dangerous predators for most marine animals. In turn, they are also tasty snacks for sharks and killer whales.

Because the photic zone is bursting with life, you will see tonnes of fish (including most of the saltwater fish we eat), sea mammals (dolphins, whales), sea turtles, and crustaceans. All of these animals do not choose to remain within their zones just to make life easier for marine biologists. Leatherback turtles, for example, live primarily in the photic zone, but they sometimes dive down into the deep sea. Some turtles have been known to swim at depths of up to 1,200 metres, two zones further down.

In this book, we focus a lot on the smallest organisms because they are so important and they are also easily overlooked. But the time has come in our underwater journey to meet the largest animal ever: the blue whale. With a maximum length of some 30 metres, this animal is larger than any dinosaur or prehistoric monster, at least as far as we know. It is not that there have not been rivals in terms of size: the megalodon, a now extinct shark that literally bears the name 'big tooth', reached lengths of up to 20 metres. It lived up to its name because, with its powerful jaws, it could easily eat a smaller whale in a few bites. Still, this shark weighed 'only' 70 tonnes, which is a lot lighter than the blue whale, which can weigh up to 199 tonnes. Until recently, people thought that the blue whale was not only the largest but also the heaviest animal ever to roam our planet. But in 2023, a fossil of a prehistoric whale, the *Perucetus colossus*, was discovered in Peru. This whale lived about 38 million years ago and is believed to have weighed a massive 340 tonnes.

How do we actually know how heavy a whale is? It is hardly practical to place the animals on a giant scale, so scientists have calculated the animal's volume based on pictures. They used these images to calculate its weight because a whale's mass density can easily be measured by taking a sample from a dead specimen and weighing it.

As impressively huge as they are, blue whales are dependent on microscopic plankton for food. They eat mainly krill, a tiny crustacean. The whales are 'batch feeders', taking in tonnes of water and then squeezing it out through their baleen plates, leaving the food behind in their mouths.

Blue whales also have an impressive communication system that allows them to communicate with each other over large distances. This is tangible proof that the ocean is not a 'silent world', as Jacques-Yves Cousteau called his famous 1956 documentary. Those grunts and clicks may be mostly inaudible to humans, but the sound levels can reach up to 180 decibels. That is louder than a jet engine. But why do we hardly hear them? The sound is at a far lower frequency than our hearing is attuned to. Whales can hear better than we do.

The blue whales are surpassed in singing skills by the humpback whale, which occasionally graces the North Sea with a visit. During mating season, the males sing complex songs consisting of both high-frequency bleeps and deep rumbles. They can even include recurring stanzas that differ per individual – as if they are singer-songwriters looking for a partner. Humpback whales are also excellent hunters: they employ a technique called bubble-net feeding, where they draw a curtain of air bubbles to trap their prey. After that, it is just a matter of opening those jaws wide and letting the baleen do its work, just like the blue whales do. Humpback whales will also slap their tails on the surface to paralyse their prey. They have a surprising number of hunting techniques.

Invisible, strange, giant marine animals have always been a source of fear. Almost every culture has myths and stories about sea monsters. Some of these are based on real animals, such as the giant squid. Sharks are also often portrayed as terrifying, although their bad reputation is undeserved. One of the main offenders in that respect is the film director Steven

Spielberg and his blockbuster *Jaws*, with its monstrous white shark in the leading role. *Jaws* further fuelled the existing fear of sharks and has made it harder for nature conservancies to create enough awareness to protect these sometimes endangered shark species. Spielberg now regrets the effect his film had on attitudes towards sharks. In a recent interview with the BBC, he admitted that *Jaws* still keeps him awake at night: 'I truly and to this day regret the decimation of the shark population because of the book and the film.' Peter Benchley, the author of the 1974 book that inspired the film of the same name, has also publicly apologised for the impact of his book on sharks and later became actively involved in nature-conservation efforts.

There is no reason for humans to be afraid of sharks. Worldwide, sharks are responsible for about 70 unprovoked attacks per year on average, of which 10 at most are lethal. While that is terribly unfortunate for the people who fall victim to these attacks, it is nothing compared to those by dogs (35,000 deadly attacks each year), snakes (60,000), or malaria mosquitos (750,000). Even cows are responsible for more fatal casualties per year (20).

Jellyfish are also more dangerous than sharks. They are responsible for some 100 deaths worldwide each year, not always directly through the poisons in their cells, but through drowning caused by extreme pain. The only truly deadly jellyfish whose toxin can kill you is the box jellyfish found in the Indo-Pacific. Others, such as the Australian box jelly or sea wasp, are highly toxic. It may not surprise you that they are found off the coast of Australia, a continent that harbours several deadly animals, including a variety of spider and snake species. Still, since 1883, only 79 fatalities resulting from a sea-wasp sting have been recorded. Heading further east, you will come across the pufferfish or fugu, a delicacy in Japan. The skin and some of the organs in this fish contain the highly toxic tetrodotoxin, a substance also found in the blue-ringed octopus and some crab and starfish species. Just a tiny quantity is fatal to humans. Chefs must clean the fish with surgical precision to remove all the toxins. When people try this themselves, things sometimes go wrong. Still, the number of fatalities is surprisingly low: on average each year, only a couple of people fall victim to pufferfish poisoning.

Let's go back to the open sea. We still have one more place to visit that speaks to the imagination: the Sargasso Sea. First of all, it is odd to call this region a sea because no land surrounds it. It is situated right in the middle of the Atlantic Ocean, roughly between the coast of North America and the Mid-Atlantic Ridge, which splits the ocean in two like a spinal column. The reason that it is called a sea is because it is a distinctive area that clearly differs from the rest of the Atlantic Ocean. The Sargasso Sea's boundaries are delimited by a network of circular ocean currents called the North Atlantic gyre, which includes the Gulf Stream. The currents turn clockwise around the Sargasso Sea, leading to the accumulation of seaweed (and plastic, unfortunately). This relatively calm sea is named after the brown algae of the genus *Sargassum*, which is prevalent in the area. This algae belongs to the same class as kelp but floats freely on the surface waters of the ocean and does not have to be attached to land. Although the Sargasso Sea has relatively little oxygen and nutrients, it is still a breeding ground for a variety of species, such as turtles and various types of fish.

It is here that we can finally unravel the mystery of the reproductive habits of eels, which many scholars like Aristotle and Freud failed to explain. It was ultimately the Danish researcher Johannes Schmidt who solved the mystery. During an expedition in the 1920s, he noted that the further he travelled into the Atlantic Ocean, the smaller the larvae of European eels (*Anguilla anguilla*) became. In the Sargasso Sea, he came across the smallest larvae ever seen. On the basis of this information, he deduced that this must be their spawning ground, although he had never actually seen the eels spawn. The eels swimming in our rivers make an incredibly long journey. They start their lives as larvae in the salty Sargasso Sea and swim via the Gulf Stream towards the brackish and fresh waters of the European rivers, where they stay and grow. Once they are big and strong enough, they return to the Sargasso Sea to spawn. They travel thousands of kilometres: the Sargasso Sea lies some 5,000 kilometres from European waters. Why so far? One interesting theory is that continental drift determined their spawning behaviour. When the continents shifted apart through plate tectonics and the Atlantic Ocean expanded, the eels simply had to swim further to reach their destination.

We still do not know all the exact details of the reproductive habits of the eel. We think that they die after their long journey westwards, but we are not sure since no one has actually seen them in action. It shows how little we know about the ocean. And that is certainly the case when it comes to life in the dark, deep sea, with its lack of oxygen, immense pressure, and life that looks like it belongs on another planet.

THE SECRETS OF THE DEEP SEA

In 2013, an extraordinary video went viral. From a submersible vehicle, a white, ghostly vision appears in front of the camera lens. As the camera zooms in, it becomes clear how huge the animal is, sparkling silver as it moves its long limbs. A giant eye, almost like the evil eye from many cultures around the Mediterranean, looks in the direction of the submersible. The scientists on board can hardly hide their excitement. It is the legendary giant squid (*Architeuthis dux*) being filmed for the first time in its natural habitat, the deep sea.

This was certainly not our first encounter with this mysterious animal. It is rumoured that in 2003, a giant squid attacked a sailor participating in the Jules Verne Trophy, a competition for the fastest circumnavigation of the world in a sailing yacht. The animal suddenly turned up between the rudder and the hull and clamped itself to the boat. The giant squid was reported to be about 10 metres long. Unfortunately, the crew failed to take pictures, leading some to refute the claim on the basis of *pictures, or it didn't happen*. The fact that the competition was named after Jules Verne was a happy coincidence because the imaginative French author wrote about a sea monster, which we would describe today as a giant squid, in his book *Twenty Thousand Leagues Under the Sea*.

A year later, proof was collected in the form of images. In 2004, the Japanese were able to capture the animal on film. This squid had a total length of over 8 metres. Prior photos of the giant squid had been taken, but they were often of animals who were dying or were not in their typical habitat.

It is plausible that the giant squid once inspired sailors to tell tall tales of the *kraken*, a mythical animal that could destroy ships with its enormous tentacles. If you exaggerate the power and size of a giant squid, you are not far off from describing a kraken, although this creature could also have been based on an octopus, the giant Pacific octopus. After a long journey at sea, it would be tempting to exaggerate your observations at the local pubs, leading to the stuff of myths and legends. On the other hand, you could say that the giant squid is living proof that you should not discount all those wild stories as hogwash: sometimes, they are based on truth, even if they are blown out of proportion.

Apart from their giant size – females can reach up to 13 metres in length – giant squids are similar to their smaller relatives, such as the common squid. They also have two tentacles for catching prey and eight shorter arms. From the stomach contents of washed-up specimens, we know that they feed on fish, smaller squid, and crustaceans. One of their biggest enemies is the sperm whale. Impressions of suckers on sperm whales indicate that the giant squid does not allow itself to be eaten by a whale without a fight. But beyond that, we do not know very much about the creature's habits.

Whatever the case, the giant squid symbolises all those deep-sea mysteries that we know so little about. We still know less about the deep sea than about the surface of the moon and, for a long time, scientists believed that there was hardly any life at those depths. We believed that life at such depths was impossible because of the immense pressure, lack of oxygen, near-freezing temperatures, and lack of sunlight. Photosynthesis is impossible, so animals have to find energy and food in some other way. Animals are mostly found in the higher levels of the sea, or they benefit from others who make the journey towards the surface. Sometimes, they can feed off the marine snow (also known as ocean dandruff) slowly drifting down to the bottom, filled with nutritious dead organisms and faeces. With a bit of luck, a dead whale may sink to the bottom, a carcass so large and nutrient-rich – the seabed equivalent of a feast – that it forms its own ecosystem.

With the lack of sunlight in the dark depths, many animals generate their own light. They do that through bioluminescence, a biological process that we have encountered before with the breathtaking sea sparkle in the North Sea. For deep-sea fish, this often involves a symbiosis with luminescent bacteria. A spectacular example is the deep-sea anglerfish, which lures prey into its open jaws with luminescent bait. You can perhaps imagine that it is not easy finding a partner in the vast darkness of the deep sea. Some types of anglerfish have developed a unique reproductive technique: the male bites into the side of the female and secretes an enzyme that fuses their skin so they become one. And somewhat less romantic: they also share their organs and blood circulation. At that point, the male only serves to provide sperm and fertilise the female. He remains fused to her until her death.

The deep sea is an inhospitable, inaccessible world. The sea floor on the edges of continents and islands is seldom flat; it slopes down. Moreover, the deep sea is home to more than 170,000 underwater mountains and 55,000 kilometres of oceanic ridges that harbour great biodiversity. There are also many crags and trenches, the most impressive of which reaches a depth of about 11 kilometres: the Mariana Trench in the Pacific Ocean. The ocean floor is, therefore, a highly diverse environment, giving rise to different ecosystems, partially determined by the depth and the terrain.

The so-called abyssal plain (*abyssos* is the Greek word for 'bottomless', 'unfathomable'), on the other hand, is flat and comprises about 76% of the seabed. These are vast plains with little relief, demarcated by oceanic ridges, mountain ranges, or trenches. They are usually covered with a thick layer of fine sediment. We do not know much about these plains, although that may soon change at locations where polymetallic nodules, potato-shaped nodules that contain manganese and other metals, are found. We will address the current discussion around the benefits and drawbacks of deep-sea mining of these nodules later. We know little about the habitat and the species that live there, but we do know that they form a biotope, or biological community, for a variety of unique species. They belong to what we call 'benthic' life, a complicated word for the bottom on (or in) which they live.

But there are more spectacular places on the sea floor, specifically on the volcanic oceanic ridges where tectonic plates diverge and seawater comes into direct contact with the magma chambers beneath. Around those geological gates of hell are what we call hydrothermal vents, hot-water vents full of minerals that bubble up from the bottom of the sea. The released minerals form chimney-shaped deposits that can reach heights of up to 25 metres: we call them 'smokers'. There are two types of smokers: black smokers, with a very high average temperature, and white smokers, which are cooler. They have both been discovered fairly recently and have probably only revealed a fraction of their secrets.

Although hydrothermal vents were spotted north of the Galapagos Islands two years prior, the smoking gun that proved the existence of black smokers came in 1979. When scientists in the submersible *Alvin* discovered black smokers on the East Pacific Rise near the Gulf of California, they could not believe their eyes. It is a dark, inhospitable environment, where the seawater temperature around the smokers can rise to 400°C. The only reason the water does not boil is because of the immense pressure. There is no light, and you will not find much nutritious sea dandruff at these depths. Not the most inhabitable of places, you would think. That is why the scientists were stunned by the incredible diversity of life they found around these smokers. They saw crabs, amphipods, shrimp, tubeworms, anemones, snails, and even an octopus aptly named *Vulcanoctopus hydrothermalis*.

The realisation soon dawned that the animals around hydrothermal vents were not dependent on the marine snow created by photosynthesis on the surface but rather derived nutrients from local chemosynthesis by special bacteria – complex chemical processes, in other words. The microorganisms that form the foundation of this food chain use hydrogen sulphide or methane as an energy source. The oxygen required for the chemical reaction reaches these zones through the thermohaline circulation, the ocean's conveyor belt we discussed in Chapter 1.

There are also white smokers, which are less hot and are driven by other chemical processes. They were not discovered until the beginning of this millennium and play a leading role in the next chapter about the origins of life.

THE WORLD IN MINIATURE

The true rulers of the sea – and land – are microbes. We can define microbes very simply as 'all creatures that we cannot see with the naked eye'. We are only now starting to realise the extent of microbial diversity, and it is truly mind-boggling. There are millions of different species; taxonomists have their work cut out for them for the next few decades. In addition to archaea and bacteria, the oldest life forms on Earth, there are countless other types of microbes. So many that it is almost impossible to divide them up into surveyable categories. Microbes are estimated to make up about 90% of all living biomass in the ocean. They are the engine of life on Earth because they produce organic substances and oxygen, and initiate or sustain essential cycles for propagating and sustaining life.

Microbes were discovered in the 17th century when the Dutchman Antonie van Leeuwenhoek carried out pioneering work with a microscope that he invented and produced. He discovered both protists and bacteria, and called them *animalcules*, 'little animals'. It must have been an incredible experience to be the first human to witness these organisms.

Recent discoveries about marine microbes, supported by developments in genetics and bioinformatics, have turned our worldview upside down. Scientists need to redraw the tree of life. We now know that higher-order species such as multi-celled animals, plants, and fungi are just a tiny fraction of all organisms on Earth, the tip of the iceberg. We owe those insights to the latest molecular technology that allows us to quickly screen what and how much life water contains. The number of bacteria, archaea, and viruses in just one litre of seawater is spectacular. It is estimated that a litre contains one billion organisms, including 10,000 types of bacteria. Of these bacteria and archaea, 99% are unknown to us: fewer than 20,000 of the estimated 10 million species have a name. Today, about 1,000 new microbes are classified and described each year. Even if biologists worked ten times faster, it would still take 1,000 years to classify and name them all.

Although the majority of these bacteria are unknown to us, you can hardly call them rare: some species are, in fact, widespread.

Prochlorococcus, for instance, was not discovered until 1987 in the Sargasso Sea, the spawning ground for eels. This tiny marine cyanobacterium is probably the most abundant photosynthetic organism on Earth. We owe a debt of gratitude to this tiny critter for all the oxygen it provides us. In the photic zone of the tropical ocean, where sunlight permeates the surface, each millilitre of surface water holds 700,000 *Prochlorococcus* cells. *Prochlorococcus* is responsible for 20% of all the oxygen produced via photosynthesis on Earth and forms the basis of the ocean's food chain. Other bacteria strains can grow at depths of up to 100–150 metres, where there is hardly any light. Together with the cyanobacterium *Synechococcus*, they are responsible for 50% of marine carbon storage. The latter bacterium was also not discovered until 1979 and is prevalent in the zones where *Prochlorococcus* does not occur.

Another example is *Pelagibacter*, the 'bacterium of the ocean', which was not isolated until 2002 and which was also discovered in the Sargasso Sea. This bacterium is possibly the most abundant organism in the ocean and even the most common bacterium on Earth.

Microbes are much more than just bacteria or archaea. Given that a microbe is simply any living organism that is not visible to the naked eye, the oxygen-releasing microalgae should also be classified as microbes. A well-known example of these microalgae are cyanobacteria, which we are more familiar with as the blue-green algae that sometimes make lakes and swimming areas unsafe for swimmers. In the chapter about pollution, we will see that it is often an excess of nutrients that causes this hazardous algal bloom.

Other important microalgae include diatoms, which account for an estimated fifth or more of the oxygen production on Earth. These diatoms belong to the protists, an odd category containing a hodgepodge of divergent organisms that we have not defined yet. The definition is a bit vague in that respect: protists are all eukaryotes (organisms whose cells contain a nucleus) that are not plants, animals, or fungi. It is the taxonomic equivalent of the drawer that you use to store all your leftover bits and bobs when you have guests coming over and do not have time to put everything back in its place. Some orderly scientists no longer use this drawer and divide the organisms up into smaller drawers. But because many people still remember this particular term from their school or university days, and because it is sometimes convenient to refer to all those diverse groups with one name, we cannot dispense with the term just yet.

The first scientist to map out the incredible diversity and beauty of protists was the German biologist and artist Ernst Haeckel. The *Protista* category was his invention, so after the above criticism on its categorisation, we do have to give credit where credit is due. His drawings from 1904 are not all that accurate from a scientific standpoint, especially given what we know today, but they are beautiful artworks that evoke a sense of wonderment for the tiniest life in the ocean. Haeckel's drawings, which included drawings of microbes, were once so popular that they influenced the Art Nouveau movement and even inspired the famous architect Antoni Gaudí. A biologist studying the tiniest organisms on Earth could not wish for more.

One group of microalgae that we could call protists are the dinoflagellates. They have nothing to do with dinosaurs: the word does not come from the Greek *deinos* (terrible) but from *dinos* (whirling). They have a twisted whip-like tail called a flagellum, which they use to move forward. A well-known example is the sea sparkle that we saw earlier in the North Sea, or at least in this chapter, and hopefully in real life as well. The zooxanthellae that live symbiotically with corals and give them their colour are also dinoflagellates.

Another noteworthy group of marine protists are the foraminifera or forams. The Greek historian Herodotus is believed to have referred to these organisms as early as the 5th century BCE when he noted that the stones of the Pyramids of Giza contained strange shapes. Later, the historian Strabo (1st century CE) suggested the theory that they were petrified lentils, the remains of the meals of labourers working on the pyramids. We now know that these were fossilised nummulites from tiny marine creatures belonging to the group foraminifera. The nummulites in the Pyramids of Giza perhaps once lived in the shallow ocean during the Eocene epoch some 56 to 34 million years ago. In comparison to their respectable geological age, the Pyramids of Giza (about 4,500 years old) are incredibly young.

The coccolithophores are also protists, but we already discussed them in our section on the white cliffs of Dover. Instead, we will take a closer look at radiolaria. They are possibly related to the foraminifera. They are tiny protists that produce their own mineral exoskeletons. Some radiolarians live in symbiosis with zooxanthellae, the algae that also have a symbiotic relationship with corals – zooxanthellae have more than one potential roommate.

You may find it hard to believe, but during the 1900 Paris Exposition, you could walk through a giant gate shaped like these microscopically small radiolarians. For his Porte Monumentale, the main entrance in Art Nouveau style, René Binet was inspired by the radiolarians from Ernst Haeckel's drawings. Haeckel himself was enthralled by the radiolarians, which he studied in Sicily. Every morning, he would go to the fish market and collect buckets of seawater teeming with tiny organisms from the fishermen. They revealed to him a 'delightful, poetic' world. At home, he would draw these creatures with feeling and precision, albeit perhaps somewhat idealised. He even included notes for artists and architects who wanted to use these natural motifs in their works. Many made thankful use of those notes.

Unfortunately, Binet's Porte Monumentale was dismantled after the exposition, as tends to happen with expo buildings.

Left: Radiolaria (Ernst Haeckel). Right: René Binet's Porte Monumentale.

We can only discuss a fraction of all marine microbes simply because most of them are still uncharted territory for science. We certainly know next to nothing about the microbes in sediments on the sea floor. Microbial cells in marine sediments are believed to make up 3% of the total of all living biomass (the sum mass of all life) on Earth. Almost 40% of those cells are archaea, of which many more are found on the continental shelves than in the open ocean.

With the latest genetic techniques, we have also discovered thousands of new species of marine viruses, even though we barely understand their role and function. An entire undeveloped scientific field, ripe for fascinating insights, lies ahead of us.

Technology has proven to be a crucial factor in accessing under-researched areas and life forms. Thanks to robots, core samples, and techniques for capturing underwater imagery, we can now access isolated environments, such as cold-water reefs, the Mid-Atlantic Ridge, submarine canyons, and the biosphere of the Earth's crust. This is how we know that organisms live at the greatest depths and even several kilometres beneath the sea floor. We are only now discovering previously unknown tiny habitats on the bottom of the sea that seem to house a distinctive, often completely new biodiversity. Hydrothermal vents, mud volcanoes, cold seeps, and brine pools are all habitats with a unique fauna that is often dependent on chemosynthesis rather than photosynthesis.

On the Mid-Atlantic Ridge, somewhere between Greenland and Norway, lies Loki's Castle. This is a field of hydrothermal vents, named after the Norwegian trickster god Loki because it was almost impossible to find. Scientists discovered a new and unique microbe species here in 2015, calling it Lokiarchaeota. These are archaea, but at the same time, they share some genes with the more complex eukaryotes, of which we are also a part. According to scientists, these Lokiarchaeota could be a missing link between the two, indicating that the eukaryotes developed from archaea.

We know so little about this mysterious microworld that we can expect fascinating times ahead. Perhaps greater insight into bacteria and archaea can also help us gain further insight into the origins of life. Everything points to the idea that life originated in the ocean, potentially in hydrothermal vents. It would also explain why the diversity there is so great. This ecosystem, safely tucked away in the ocean's womb, has had plenty of time to take on all sorts of forms, often much more exotic than we are used to on land. It is worth taking a closer look at the origins of all those fascinating life forms.

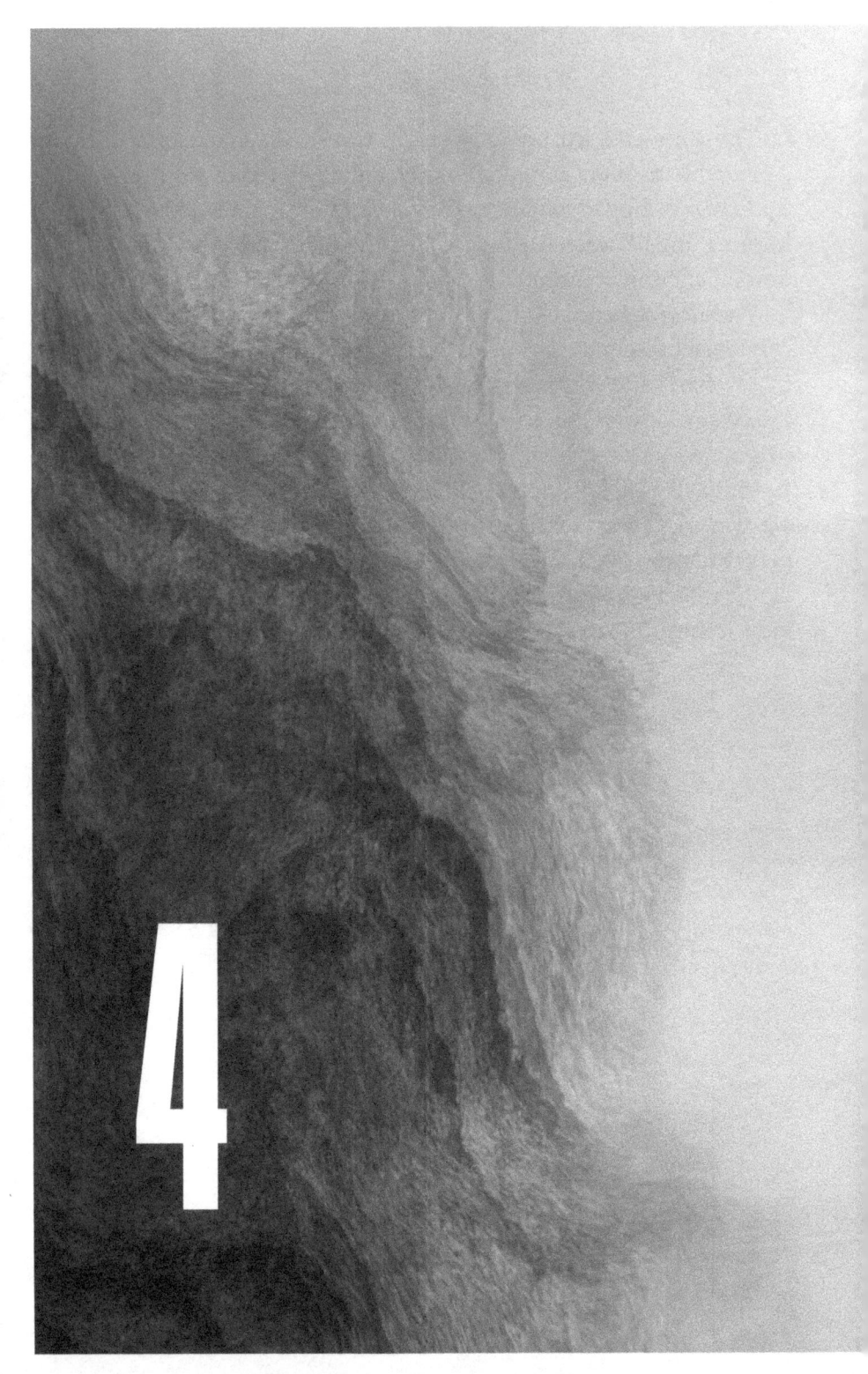

THE OCEAN, MOTHER OF ALL LIFE

> "Overhead the albatross
> Hangs motionless upon the air
> And deep beneath the rolling waves
> In labyrinths of coral caves
> The echo of a distant time
> Comes willowing across the sand
> And everything is green and submarine"
> **PINK FLOYD ('ECHOES')**

Where do we come from? How did life originate? People have been asking themselves this question ever since cultures evolved and humans have had time to think about more than just purely practical problems. Mythology, religion, philosophy, and science have all tried to answer these questions. Theories abound about the origins of life: it is the result of divine creation, an extraterrestrial invasion or bombardment with microbes, spontaneous generation (Aristotle's idea), a chemical primordial soup, and many more.

Most modern theories assume that life must have originated in water in some way. That is why our search for extraterrestrial life has become a search for water on other planets or moons. And since life on Earth may have originated in the ocean, and because of our timeless curiosity about our origins and, more recently, life beyond our planet, it deserves a chapter in this book. In order to understand the immense diversity of marine life, we need to go back to its origins. The more knowledge we have about the origins of life, the better we can understand how the Earth grew to become a habitable place for us as humans.

First disclaimer: we still do not know exactly how life on Earth began. What we do know is based on hypotheses, which we may someday end up refuting. But we cannot underestimate the advances that have been made in science in recent years. Today, we can establish a plausible theory that aligns with recent insights from the various sciences that are needed to reconstruct a coherent narrative.

Second disclaimer: this part of the book is a difficult read. We cannot expect to answer such a complex question with a brief, straightforward answer. So feel free to skip this chapter if you prefer some light reading. But if you are willing to make the effort, you will hopefully be rewarded with a glimpse into one of the most fascinating mysteries of our existence. In any case, we have to start our journey of discovery with the birth of our planet and its ocean before life ever existed.

THE YOUNG EARTH AND THE PREHISTORIC OCEAN

The universe, the cosmic party to which we are temporary and very late guests, started some 13.8 billion years ago with a big bang. It was the Belgian priest and scientist Georges Lemaître who first described the Big Bang in 1931, although he formulated it differently. To understand how our Earth originated, we need to fast-forward a couple of billion years. About 4.56 billion years ago, a fast-spinning cloud of gas and dust orbited the sun, our sun. One of those colliding clumps of rock in that cloud eventually became the planet Earth, little more than a spinning orb of magma at the time. After several million years, the Earth had developed a core composed of heavy metals and surrounded by a lighter mantle. Shortly afterwards, a new spectacular event took place: Theia, a small planet about the size of Mars, named after one of the Titans from Greek mythology, collided with the young Earth. It sounds like the ultimate apocalyptic event from the film *Melancholia* by Lars von Trier, in which the planet Melancholia heads straight for our Earth. But this actually really happened. Although, thankfully, no one was around to witness it at the time. One of the fragments from the cosmic collision between the Earth and Theia became our old friend, the moon. In Greek mythology, Theia is the mother of the moon goddess (Selene), drawing a parallel between the myth and the birth of our moon.

It is difficult to reconstruct what the world looked like four billion years ago, but we believe that our planet started as a scorching hell. Magma gushed over the unstable surface and cooled down, and silicate (a mineral) solidified into solid rock on the surface. It will not surprise you that geologists refer to this phase as the Hadean period, after the Greek god of the underworld. Meanwhile, other minerals started to form the Earth's crust. Gases released by volcanic eruptions and meteor impacts led to the development of a volatile early atmosphere consisting of gases such as water vapour, carbon dioxide, sulphur dioxide, nitrogen gas, and nitrous oxides.

At some point, when temperatures had dropped enough, all the water vapour turned liquid and led to an endless period of rain, like an endlessly rainy November day, but without a soul or warm café in sight. There was no land back then, either. Quite the opposite: a giant ocean covering the entire Earth's surface started to form. It probably took some 10,000 years – the geological equivalent of a blink of an eye – to fill the ocean with a volume of water twice that of our current ocean. There are no indications that there was any land jutting out of the water; there were no continents. It was a wondrous water world, a worldwide prehistoric ocean that covered the world from about 4.3 to 3.2 billion years ago.

Where did all that water come from? Opinions are divided in this respect. In any case, there was a lot of water vapour in the atmosphere from the particle cloud travelling around the sun, which literally precipitated as temperatures dropped. However, many suspect that ice meteorites may also have supplied the necessary water.

Whatever the reason, the Earth was one giant 'blue planet', much more so than it is now. On the one hand, it was a peaceful water world with a sea floor that lay many kilometres beneath the surface. But what happened above the water's surface could hardly be considered calm. Conditions were rough, to say the least. The days were shorter because the Earth was spinning twice as fast around its axis at the time, the ultraviolet radiation was devastating due to the lack of an ozone layer, the tides were twenty times stronger because of the proximity of the moon, and meteorites bombarded the planet. We may sometimes complain about the weather but, back then, temperatures fluctuated between freezing and 100°C, and the weather was such a chaotic maelstrom that you would not even know where to start complaining. Not that you would have been able to draw breath to complain in the first place because there was no oxygen in the atmosphere. Oxygen did not enter the atmosphere until the revolutionary development of photosynthesis some 2.4 billion years ago. But, even back then, seawater was already briny: vapours containing sodium chloride precipitated and dissolved in the seawater. There was no fresh water at the time. The ocean was a more stable warm temperature. With a pH of 5 to 6, it was more acidic than it is today at pH 8, and it was very rich in CO_2. Plate tectonics did not exist yet, and underwater volcanoes were a common phenomenon. Hydrothermal activity in the prehistoric ocean was much more violent because the temperature of the Earth's mantle was 300°C higher than it is today.

Now that we have set the scene, the question remains: how did life originate? To address that question, we need to look at the best hypothesis. We are not going to discuss all the theories that have been suggested to explain life on Earth. However, one of them is deeply ingrained in our collective consciousness, although it is becoming less of a plausible explanation today. We first need to get this hypothesis out of the way before we can move on and suggest a better alternative.

THE PRIMORDIAL-SOUP HYPOTHESIS

Anyone who has studied the sciences in the distant past may remember that life originated from a sort of primordial soup. The idea behind it was of biblical, apocalyptic proportions:

shafts of lightning light up the dark, lifeless planet while countless volcanoes erupt and turn the sea into a boiling, writhing mass. This is an inferno that makes Dante's look like a haven of tranquillity. Meanwhile, organic molecules and other building blocks float around and concentrate in warm pools on land, ready for the next step in their chemical evolution. A barrage of lightning strikes then lights the fuse of life.

This theory has several variations, not all as spectacular as the one described above. The primordial soup theory builds on Darwin's hypothesis of 'warm little ponds', which he believed were the birthplace of life. In the 1920s, the biologists Alexander Oparin and John Haldane proposed the idea of a primordial soup. Later, a world-famous experiment by chemists Stanley Miller and Nobel laureate Harold Urey supported this theory. In 1953, Miller and Urey succeeded in synthesising amino acids, the building blocks of proteins, in a laboratory environment. They did so by replicating lightning strikes in a closed system of water and a mixture of gases (methane, ammonia, nitrogen, and hydrogen), which they believed replicated the primordial atmosphere. According to the hypothesis, the accumulation of these precursors to life in a water-rich environment would eventually lead to that fertile primordial soup. Many subsequent experiments proved that organic building blocks could be developed from inorganic substances.

All this sounds great, but there are severe limitations to the primordial-soup theory, which was generally accepted until very recently. To begin with, the assumed composition of the atmosphere was not correct. It was not methane, ammonia, and hydrogen (the composition of Jupiter's atmosphere) that formed the atmosphere of young Earth, but the relatively inert carbon dioxide (CO_2), nitrogen gas (N_2), and water vapour.

Moreover, scientists date the origins of life to about four billion years ago. But the planet would probably have been entirely covered by water back then. There was no dry land and, therefore, no shallow ponds containing primordial soup. Besides, pools and ponds are highly unstable, transient systems. There was no protective ozone layer, so the powerful ultraviolet radiation from the sun would have killed any primitive life instantly. Therefore, we should nuance the apocalyptic imagery of the young Earth as a volcanic hell with a fiery ocean: beneath the protective ocean surface, the Earth of four billion years ago was a relatively calm but oxygen-deprived underwater world devoid of life, although that was soon about to change.

We have yet to deliver the final death blow to the primordialsoup theory, but the arguments stated above make it less plausible. Nonetheless, the theory, and particularly Miller and Urey's experiment, proved that it is possible to make organic molecules from inorganic material under the right circumstances. These inorganic materials could have provided useful building blocks to the ocean. But given the current science on the subject, life probably did not occur in Darwin's warm little ponds.

BLACK AND WHITE SMOKERS

For two generations, the primordial-soup theory had the scientific community in its grip. It was not until the 1970s that the discovery of the black smokers – hydrothermal vents deep in the ocean – turned this branch of science upside down.

Black smokers owe their name to their chimney-like shapes and the black 'smoke' they spew, filled with dissolved minerals, carbon, and energy. Not surprisingly, some started to wonder whether these were the locations where life could once have originated. Still, some factors could be used as arguments against smokers being the birthplace of life on Earth. Can new life thrive at extreme temperatures of 250 to 400°C? And what about their explosive nature and short life span? Are these black smokers suitably stable environments for establishing life on Earth? Science had once more reached a dead end in its search for the origins of life.

Still, a new avenue of research opened up when, in 2000, a team of researchers led by marine geologist Deborah Kelley discovered a different type of hydrothermal vent in the middle of the Atlantic Ocean: alkaline hydrothermal vents or 'white smokers'. These smokers owe their white colour to the precipitation of calcium carbonate. Situated slightly away from the mid-oceanic ridges, they tower some 60 metres above the sea floor. The system discovered in the middle of the Atlantic Ocean was dubbed 'the Lost City', referring to the other-worldly and desolate feeling the scientists experienced upon seeing these underwater images.

The white smokers were a much better candidate for being the cradle of life. In contrast to the black smokers, which were characterised by their short-lived nature and extreme circumstances, the white varieties are more stable, with a temperature suitable for the friendly chemistry required for the emergence of life. Whereas black smokers were the product of the brutish violence caused when magma comes into contact with seawater, white smokers are the result of a much more subtle process. They are the products of a chemical reaction between water and the hard rock in the Earth's mantle, which is rich in a light-green mineral called olivine. When olivine comes into contact with water, it reacts to form the mineral serpentine, plus large amounts of hydrogen gas. The gas dissolves in the warm alkaline liquid and bubbles up towards the seabed, where it cools and precipitates as calcium carbonate, the stuff that makes those beautiful towers. Unlike the black smokers, those towers are not open chimneys. They look more like sponges: a labyrinth of minuscule pores separated by thin walls and saturated with alkaline fluids. The structure of white smokers is just the type of labyrinth that would be necessary to establish life: they look like giant cells (in the biological sense of the word).

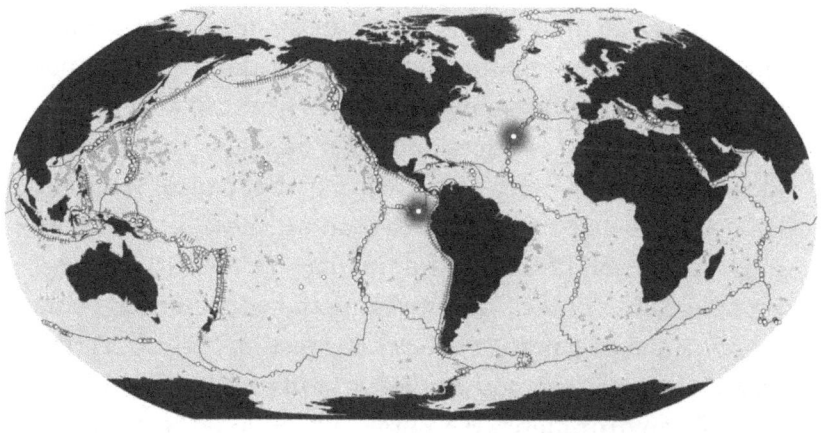

Figure 4. Hydrothermal vents (white dots) are located primarily on ocean ridges, where tectonic plates move apart and where volcanic activity often takes place. The large dot to the left in the Pacific Ocean is where the first black smoker was discovered; the large dot on the right in the Atlantic Ocean is where the white smokers of the Lost City were found. To date, 721 hydrothermal vents have been mapped (Beaulieu & Szafranski, 2020).

All in all, those white smokers, which were most probably around when the Earth was still young, seem to be a good candidate for the birthplace of life. There are sufficient minerals and organic molecules, there is chemical energy, and the temperature seems about right. You do not need sunlight, lightning storms, volcanic eruptions, or other violent events to provide an energy supply. White smokers have the potential to concentrate organic molecules, thanks to their compartmentalisation into cells and the presence of what we call gradients, which are gradual changes in various factors (including redox reactions for the exchange of electrons, temperature, density, and acidity). We will explain the crucial role of these gradients later.

What is also important is that white smokers offer a stable environment. These pore-riddled towers and the surrounding ocean protect vulnerable molecules from radiation, dehydration, meteorite impacts, and other dangers. They provide the proper circumstances for the complex chemical reactions needed to set cell metabolism in motion. These reactions are universal for all life, from the smallest bacteria to the largest whale. The idea of life emerging in hydrothermal vents is confirmed by genetic analyses suggesting that LUCA (the Last Universal Common Ancestor of life on Earth) lived in such an environment. We know this by looking at which genes the most primitive archaea and bacteria have in common; those genes must have already been present in their common ancestor and reveal to us in which type of environment they evolved.

But to truly understand how life originated, we need to ask ourselves a deceptively simple question first: what *is* life?

THE FIRST LIFE

The oldest signs of life that we have discovered thus far were found in prehistoric rock layers in Greenland. Those are about 3.8 billion years old. It is not a fossilised bacteria or an imprint but just plain graphite, a form of carbon that you also find in pencils. This graphite may be biological in nature, as suggested by its isotopes (isotopes are variations of an element with a different number of neutrons in their nuclei). However, whether this is actual proof is still up for debate. Although we generally tend to make a clear distinction between rocks (geology) and life (biology), that line is not as clearly defined when it comes to early life forms.

We find a more concrete indication of life in the stromatolites, rock structures formed by cyanobacteria and other bacteria. In rocks that are 3.5 billion years old in Australia and South Africa, we find microfossils that strongly resemble cells, as well as deposits that are reminiscent of stromatolites. That tiny creatures can turn into rocks should not surprise us because earlier in this book we have already seen a similar process where tiny sea creatures form calcium carbonate deposits.

So, life developed very early on after the Earth was born, most probably as early as between 4.3 and 3.8 billion years ago. Between 3.5 and 3.2 billion years ago, bacteria developed the main metabolism mechanisms (such as breathing and photosynthesis). And all the essential nutrient cycles (carbon, nitrogen, sulphur, iron, and so on) were already active 2.5 billion years ago.

We can, therefore, say that life on Earth developed quite quickly after the Earth was formed. The Belgian biochemist and Nobel laureate Christian de Duve states that this was bound to happen because of the laws of chemistry, which hold that a chemical reaction usually occurs quickly or not at all. If a reaction takes thousands of years to complete, chances are that the substances involved in the reaction disappear or break down unless other, quicker reactions constantly replenish them. So, on the basis of that logic, we can say that life must have developed quickly rather than over the span of millions of years.

But what do we mean when we talk about 'life'? Until now, we have been avoiding this tricky issue because it is not easy to draw up a clear definition. One of the most commonly used definitions is the one adopted by the space agency NASA: "Life is a self-sustaining chemical system capable of Darwinian evolution." The weakness in this definition lies in the word 'self-sustaining' because life does not strictly have to be self-sustaining. Life is a highly unstable condition requiring a constant supply of energy and materials. Just think of all the food that we have to eat every day. Life is more of an open system that exchanges matter and energy with its surroundings to maintain its state, which is far removed from a state of equilibrium.

Most modern definitions, therefore, avoid the noun 'life' and instead use the adjective 'living', which points to a changeable state: living beings are autonomous systems with open-ended evolution capabilities. The different definitions of life each focus on one or more characteristics of living beings, such as their ability to reproduce or compartmentalise (into cells or organs). Those different characteristics are traditionally studied in different scientific disciplines which historically have always operated independently. Research into the origins of life is a broad scientific field that requires an interdisciplinary approach and expertise. Sadly, there are few scientists today who adopt such an approach.

For those of you whom we have not convinced yet, let's be clear: life is exceptionally complex. So let's take a practical approach: 'to live' is a verb, and the four most important characteristics that all living beings share is that they (1) are capable of reproducing, (2) are in a state far removed from thermodynamic equilibrium (from an entropy standpoint: a constant supply of energy is required to maintain their state), (3) have some form of compartmentalisation (think cells and membranes), and (4) have the potential to evolve.

BACK TO THE (HYDROTHERMAL) SOURCE

We can now try to tie up any loose ends. The hydrothermal vents that scientists discovered in the last century were also present in the prehistoric ocean four billion years ago. The black and white smokers would probably have been slightly different in those days because there was no oxygen and the CO_2 levels were about 100 to 1,000 times higher than they are today. Oxygen was catastrophic for early life on Earth, a deadly poison for the first organisms.

The secret of hydrothermal vents is that they have gradients, which we know as gradual changes in different characteristics: acidity, temperature, density, ion concentrations, and so on. The greater the variation, the greater the chance that useful chemical reactions take place. The first step towards life is what we call 'prebiotic' (literally meaning 'before life') chemistry, in which complex molecules were created. According to our current hypothesis, these chemical reactions took place near the hydrothermal vents in the ocean.

There must have been a considerable difference between the cool, acidic ocean and the alkaline, hydrogen-rich liquids coming from the hydrothermal vents. In technical terms, it concerns not only a difference in temperature but also in acidity and, therefore, in the number of protons: hydrogen ions (H^+) make the ocean more acidic. What we call 'acid' is nothing more than a substance that releases hydrogen ions (H^+) when dissolved in water.

The situation in the prehistoric ocean would have been stable for at least 100,000 years, enough time to create the complex molecules required for life. However, for life to develop, the circumstances must have been, on the one hand, stable and favourable enough for longer periods of time and, on the other, dynamic enough to make possible prebiotic chemistry and its gradual evolution into biology.

One last potential key characteristic of hydrothermal vents is their mineral deposits, namely the 'green rust' and various iron sulphide compounds. Those minerals may have helped shape organic substances and could even have been the precursors to cells with their layered structures. Water is trapped between those layers, and the gradients would have caused high concentrations of prebiotic products (organic molecules such as proteins and lipids) in those layers. And thus life begins. Eventually, the first common ancestor escapes into the ocean – either from the green-rust layers or the porous white smokers – and the rest, as they say, is... natural selection.

Granted, the chemistry behind it is very complex, and we do not know all the details. But for now, the pieces seem to fit. And we can exclude almost all other environments on Earth as realistic options for the origins of life, simply because they do not meet all the requirements like hydrothermal vents do.

HOW MARINE LIFE CHANGED THE PLANET

Regardless of how life began, as soon as the prebiotic phase of complex chemistry was complete, the ocean contained two very different life forms: archaea and bacteria. We call them prokaryotes because they do not have a cell nucleus (the Greek *karyon* means nucleus). Now, let's travel some time into the future. The Proterozoic aeon, starting 2.5 billion and ending 539 million years ago, is characterised by plate tectonics and the development of an oxygen-rich atmosphere. In the beginning, there was hardly any oxygen. Photosynthesis literally injected oxygen into the ocean and the atmosphere. It is in this new, stable condition that complex life as we know it evolved.

Complex life forms with a nucleus, the eukaryotes, were the one-off product of the symbiosis between an archaeon (the singular form of archaea) and a bacterium that turned into a mitochondrion, the cell's energy powerhouse. Eukaryotes developed some 1.7 billion years ago but did not thrive until about 800 million years ago. This interim period between 1.8 and 0.8 billion years ago is known as the Boring Billion or the Earth's Middle Ages. The mountains stopped growing, existing mountains even levelled off thanks to gravity, and the simple life forms in the ocean evolved incredibly slowly. It is possible that the flattened continents provided fewer minerals, such as phosphorous, to feed the ocean and therefore slowed down the development of complex life. In other words, the early life forms starved due to a lack of phosphorous and other essential elements, drastically decreasing their productivity and, hence, their evolutionary rate.

Starting 2.4 billion years ago, an atmospheric revolution took place, one of the most critical events in Earth's history:

the Great Oxidation Event (GOE). Lynn Margulis referred to it slightly less tactfully as the Oxygen Holocaust. Many organisms died during this event because oxygen is toxic to many organisms, even today. We suspect that the oxygen came from cyanobacteria, the blue-green algae. Later on, other events would take place that would increase oxygen concentrations in the atmosphere, but we do not want to get ahead of ourselves. About 500 million years ago, during the Cambrian explosion of life, oxygen levels were about the same as they are today.

Oxygen provides energy to life (by burning food to create carbon dioxide and water) and allows it to grow into big and complex forms. Some bacteria can live without oxygen, but their reactions provide far less energy. Only oxygen provides enough energy to meet the needs of multicellular life. Without it, there would only be microscopic life on Earth.

The rest is history (or palaeontology, to be more precise). The complex eukaryotic cell, sex as a means of reproduction, focused movement (through mechanisms such as muscles), and sight all developed in the ocean. Sight, by the way, is a sense only found in animals. The first eyes with lenses were found in the extinct trilobites, a successful group of arthropods (which includes crustaceans) that inhabited the ocean between 521 and 250 million years ago.

Time for evolution to take place is definitely a factor that made possible the extensive diversity in marine life. Life originated in the sea, so it had plenty of time to evolve. The colonisation of land is, in fact, a relatively recent development. It was not until 500 million years ago that plants and animals moved onto dry land and that plants gave us our green continents.

EXTRATERRESTRIAL OCEANS, EXTRATERRESTRIAL LIFE?

The above facts, hypotheses, and assumptions raise several questions. There are other oceans in our universe, even in our solar system. What does that mean for extraterrestrial life? Is life possible in other places in our solar system and beyond?

By now we know that a handful of chemical elements and alkaline hydrothermal vents were sufficient to generate life. Microbial life on Earth developed 'easily'. Although we are not clear on the details yet, we know what was needed and roughly which mechanisms were involved. These processes led to the creation of not one but two different life forms: the archaea and the bacteria. Chances are that a similar process took place elsewhere in the universe. It is even conceivable that this also happened elsewhere within our own solar system.

There are several arguments for extraterrestrial life. Elements such as hydrogen and helium spread out relatively evenly across the universe following the Big Bang. The same goes for elements such as carbon, oxygen, nitrogen, and other elements that developed at a later stage, often through the nuclear reactions in stars. There is no reason to believe that their concentrations would be higher in our solar system than elsewhere. Water in our trusted H_2O form is even present in large areas throughout the universe because it is easily formed in interstellar clouds and found in 'rings' around planets and moons. The rings around Saturn contain lots of ice, for instance. So, from a purely statistical standpoint, there is a considerable chance that, somewhere out there, the right circumstances arose in which those elements led to the emergence of life. There are countless star systems. Moreover, we have seen that life is very resilient and can exist in even the most extreme of circumstances.

But there is more. Alkaline hydrothermal vents are probably common on wet rocky planets or moons elsewhere in the universe. The icy moons Europa (orbiting Jupiter) and Enceladus (orbiting Saturn) are good candidates in our solar system. Their surfaces are barren (the sun's ultraviolet radiation kills all life on the surface) but, beneath the thick icy crust, they have an ocean resting on a rocky seabed, which makes possible direct interaction between water and rocks. Hydrothermal activity there could be a result of either underwater volcanic activity or the energy created by incredibly powerful tidal forces. Those factors make these celestial bodies fascinating subjects in our search for extraterrestrial life.

Considering all of this, we can safely assume that life (similar to the bacteria and archaea that we know on Earth) can be found throughout the universe.

The Swiss astronomer Didier Queloz, who received the Nobel prize in 2019 for his discovery of the very first exoplanet in 1995, believes in extraterrestrial life. He predicts that within the next 30 years, we can develop the technology to trace biochemical indicators of life on an exoplanet from space. 'I am confident that humans will have detected alien life in 100 years' time,' Queloz has stated in several interviews.

MASS EXTINCTIONS AND THE ROBUST OCEAN

Once life had established a foothold on Earth, it was almost impossible to get rid of it. As stated by the character Malcolm, played by Jeff Goldblum, in the film *Jurassic Park*: 'Life finds a way.' Marine heat waves or heavy bombardments with space rubble were not able to bring life to its knees. Once microorganisms show up somewhere, they are there to stay. Even during mass extinctions, there are always habitats that serve as reservoirs for life. That does not necessarily apply to us as humans: when the next mass extinction event takes place, there is a pretty good chance that we will go down with the ship, but the robust microbes will once again survive.

In the past 550 million years, five major mass extinction events have taken place, marking the boundaries of several major geological periods. The extinctions were always the result of climate change caused by rising or falling atmospheric carbon-dioxide levels, and of ocean acidification. We will see in the chapter about climate change that there are parallels with the current climate crisis.

The first mass extinction event was in the Late Ordovician period 445 million years ago, when 85% of all marine life disappeared. Volcanic activity probably led to rising greenhouse-gas levels, leading to global warming and a lack of oxygen in the water. The subsequent precipitation of volcanic ash led to an ice age lasting a million years.

The second multi-staged event took place in the late Devonian period, 372 million years ago. We do not know precisely what caused it. There may have been heavy meteorite strikes, but it could also have resulted from the sudden abundance of land-based plants and their rapid photosynthesis. Too much oxygen and a CO_2 shortage, in combination with a decrease in the Earth's temperature, could have been the death blow for many species. There is no real 'ideal' balance between oxygen and CO_2: it depends on the condition of life at that moment. This complex mass extinction event lasted 13 million years, with 70% of all species on Earth disappearing.

The third mass extinction event occurred about 251 million years ago, during the transition between the Permian and the Triassic periods – the transition to the epoch of the dinosaurs. In this event, 70% of all land animals and 80% of all marine animals disappeared. At the time, the Pangaea supercontinent and the Panthalassa superocean covered the Earth. The event was triggered by a million years of violent volcanic activity in the area that we now know as Siberia. Carbon-dioxide concentrations in the atmosphere increased strongly, temperatures rose, and the ocean acidified. The trilobites, fascinating creatures that we now only know from fossils, became extinct. After this particular episode, the Earth was very thinly populated for five million years. It is from these 'survivors' that the biodiversity we know today evolved.

The fourth mass extinction event happened 201 million years ago during the transition from the Triassic to the Jurassic periods. It was a catastrophic event in which 75% of all species disappeared in less than 10,000 years. The probable cause was the breaking up of the Pangaea continent, which came – once again – with increased volcanic activity and rising CO_2 levels in the atmosphere.

The fifth and most famous mass extinction is the one that killed the dinosaurs. This event occurred 66 million years ago when a comet struck the planet near Yucatan in Mexico. It is no coincidence that this marks the end of the Cretaceous period: we demarcated this period in hindsight, using the meteor impact as an end date. Half of all marine life ceased to exist and, as we learned at school, the dinosaurs became extinct. All dinosaurs? Not all of them; a tiny group survived. We call them birds. Not only did birds flourish after the meteor strike, but also mammals and, eventually, humans.

Many mass extinction events were the result of climate change involving greenhouse gases. It is a warning for times to come: the current favourable circumstances on our planet could suddenly and drastically change. Today, we are faced with the threat of a new, sixth mass extinction event due to climate change caused by humans. Some say that it has already begun.

We humans have also been responsible for the extinction of quite a lot of species in the past. Although most marine life still has to be mapped out, we do know of several marine-species extinctions, mostly of birds, mammals, and fish that spent part of their life cycle on land and came into contact with humans. The earliest examples that we know of are several seabird species, including Olson's petrel and the Saint Helena petrel, shortly after the island of Saint Helena was discovered in 1502. In 1768, 27 years after Europeans discovered it in Siberia, a marine mammal called the Steller's sea cow was hunted to extinction for its meat and fat. The great auk, which bore some superficial similarities to the penguin, could once be found on European shores. On 3 July 1844 on Eldey, near Iceland, the last couple was killed and the last egg collected. The last spectacled cormorant of Bering Island was sighted in 1850, and the Christmas sandpiper native to Kiribati disappeared after 1850, probably exterminated by the rats and pets that humans had brought with them. The labrador duck was last seen on 12 December 1878. It was wiped out because it was hunted for its meat and feathers, but egg harvesting and food shortages caused by the overexploitation of mussels could also have played a role.

The North American sea mink, a member of the mustelid family, died out in 1894, hunted to extinction for its fur. The last New Zealand grayling (a once abundant fish known to the

Māori as 'upokororo') was spotted in 1923. The first marine teleost we lost was the smooth handfish (*Sympterichthys unipennis*), a bizarre fish from Tasmania with fins shaped like hands. The Canary Islands oystercatcher, presumably a subspecies of the Eurasian oystercatcher, became extinct around 1940. The Caribbean monk seal followed in 1952. The Japanese sea lion was last caught in 1974. Later sightings have been unconfirmed; this species is considered extinct.

There are also examples of recently extinct invertebrates: the eelgrass limpet *Lottia alveus* was a common sight on the east coast of North America until the late 1920s. A sudden decline in fields of common seagrass (*Zostera marina*), decimated by a slime mould, led to its extinction, which was not noticed until much later.

Despite all these extinction events, life nevertheless continues to be relatively robust, especially in the ocean. This robustness accounts for the large diversity of ocean. Particularly during mass extinction events, the marine fauna proved better insulated from extinction, which made it possible for diversity to be maintained and to increase after the catastrophic event, both in terms of species diversity and ecological roles.

This is actually cause for solace and hope. Life is robust, as are its most important processes. Life will not disappear following catastrophic events, and life in the ocean will continue to exist – at least, as long as the Earth itself remains. Somehow this is a comforting thought: even if life on land were to disappear, we would always have a reservoir in the ocean. Major mass extinction events have not led to the extermination of life. The oceanographer Paul Falkowski claims that a handful of chemical reactions and, therefore, just a couple of thousand genes are sufficient to sustain life and the biogeochemical cycles. But this comforting message does not necessarily apply to us as humans: we are less robust. Then again, no hope or comfort will be needed if we are no longer around to feel it.

So, the ocean may save life when the planet is going through hard times, but that does not necessarily include us. Thankfully, we can focus on the treasures and benefits that the ocean does still provide us today. And these turn out to be surprisingly numerous.

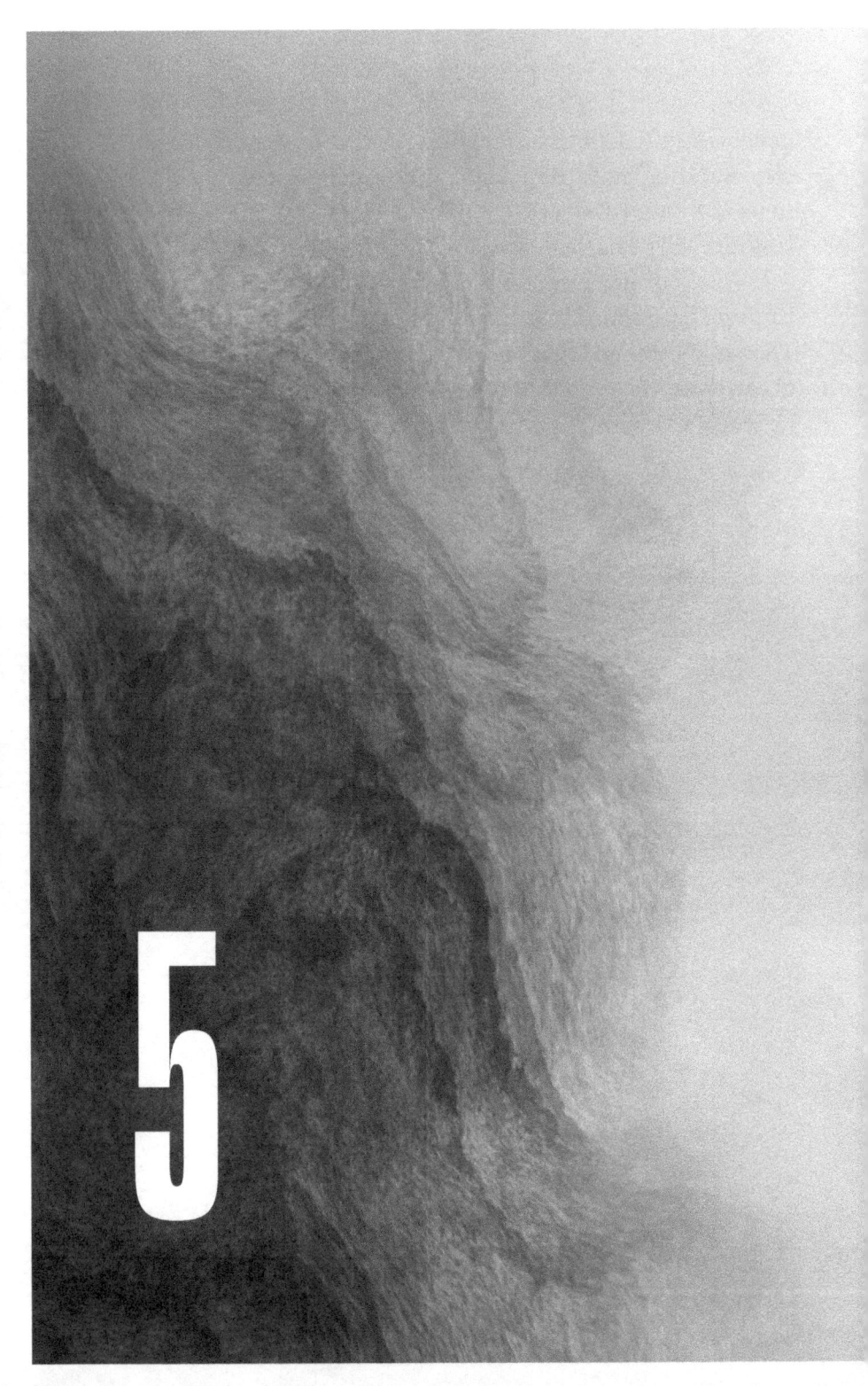

THE INVISIBLE DOCTOR AND THERAPIST

> "Smell the sea and feel the sky.
> Let your soul and spirit fly."
> **VAN MORRISON ('INTO THE MYSTIC')**

People go to the shore to catch a breath of fresh air, find a bit of peace and quiet, or just to have fun. The immeasurable expanse of the sea evokes a sense of freedom and vastness, perhaps even amazement, especially in combination with the interplay of light on the water's surface. The tumbling and shimmering crests of the foamy waves are timeless and yet constantly changing. If you swim in the surf and let the waves wash over you, you will be feeling a natural force millions of years old – almost as old as the Earth itself. The cold water may startle you at first, but you often emerge feeling reborn.

You do not have to be a scientist to sense that the sea has a restorative influence on people, but a sense or feeling does not constitute proof. So, it is worth exploring what scientific research has to say about the curative properties of the sea; studies sometimes confirm what we already instinctively know, but they have also led to the discovery of several surprising new insights. In this chapter, we will focus on the often unknown health benefits of 'Doctor Sea'.

ANCIENT WISDOM

We now know that Aristotle was one of the first people to observe marine life systematically. However, when it comes to the sea's health benefits, he was preceded by several other scholars who deserve at least as much attention. Unfortunately, many are long forgotten because their names were never recorded, their writings have been lost to history, or because the knowledge came from several people instead of a single individual. There are indications, for instance, that the ancient Egyptians adopted sea therapy practices.

Perhaps it should not be surprising that we find the first concrete indications of the healing power of seawater among the ancient Greeks. As far as we know, the philosopher and medical writer Alcmaeon of Croton (approx. 470 BCE) was the first man to link human health to water quality. His name probably does not ring any bells, but he was one of the great pre-Socratic thinkers, as we today somewhat irreverently refer to the Greek thinkers before Socrates. Alcmaeon was probably the first philosopher to consider our brains as the seat of our intellect, and his main contributions to early science were in the field of medicine. Just like other medical practitioners in those days, he thought that illness resulted from an imbalance between contrasting properties known as 'humours', such as hot and cold, wet and dry, bitter and sweet, and so on. And, according to Alcmaeon, illness could also be caused by poor water quality.

Although many influential scholars such as Alcmaeon preceded him, Hippocrates is generally known as the 'father of medicine'. Even today, young physicians are often still required to take the Hippocratic Oath, which the Romans summarised in Latin as *primum non nocere:* first, do no harm. Hippocrates further developed Alcmaeon's theory about water quality, explaining the connection between place of residence and the presence of illnesses in his treatise titled 'On Airs, Waters and Places' (approx. 400 BCE). Proximity to the sea and climate were considered important health factors in those days. Hippocrates also suggested that breathing in steam from (boiling) seawater could be restorative to humans. What he could not have known is that heating the water forms air bubbles that rise to the surface and vaporise into minuscule drops called aerosols. In any case, he and his suggestions were far ahead of his time. Several years ago, researchers from Ghent University were able to prove that marine aerosols, tiny particles that form in the air through waves and wind, contain organic substances that could be beneficial to humans. We will come back to that later.

Doctors breathed new life into the idea of marine therapy in the 18th and 19th centuries, first in England and later in continental Europe. In Belgium, various doctors described the health benefits of the sea and tried to apply these benefits in practice. In 1843, the Belgian doctor Louis Verhaeghe wrote *Les bains de mer d'Ostende, leurs effets physiologiques et thérapeutiques* (Ostend's Sea Baths, their physiological and therapeutic effects). It was a fervent argument in favour of the exceptional healing properties of Ostend's beach as well as the seawater, and thus of sea air and sunlight in general. Verhaeghe searched for the characteristics that made the sea air beneficial, but he was honest enough to state that 'this question was a long way from having a definitive answer'. He was more definite in his statements about the Ostend seawater, which he believed could help remedy digestive problems, respiratory tract infections, heart conditions, rheumatism, epilepsy, and feminine infertility. He tried to support and prove his statements with case studies, but the scientific evidence was relatively meagre. He did promote Ostend in the process: the upper class spent their holidays there, and some even spent weeks following treatments at the spa.

The term 'thalassotherapy' became popular from 1865 onwards. This term is derived from the Greek words *thalassa* (sea) and *therapia* (treatment), and refers to the use of seawater, marine products such as algae extracts, and the coastal climate and its air as a form of therapy. Towards the end of the 19th century and in the early 20th century, the popularity of thalassotherapy led to the establishment of health centres and marine sanatoria along the coasts of the Atlantic Ocean and the North Sea, places to get away from the polluted air in the cities. Those institutions often had large patios or balconies where patients with conditions such as tuberculosis and rheumatism could benefit from the sunlight and the sea air.

While the coast continued to draw tourists for holidays and relaxation, especially during the interwar period, the idea that sea air and water could be beneficial to our general health also persisted in medical circles.

MODERN RESEARCH: THE BLUE GYM, OLD FRIENDS, AND SEA AIR

As we can see, there is a rich, long-standing tradition of general wisdom regarding the health benefits of the sea, but there is usually little scientific evidence to support these ideas. Gradually, however, several of these kernels of ancient wisdom are now being confirmed by modern science. The realisation that there is a correlation between the sea and human health has even led to a new interdisciplinary research field in the past decade: Oceans and Human Health (OHH). This field addresses topics such as the vulnerability of ecosystems and the consequences of their collapse for humans, but also the more positive effects of the sea on our health. We have known for quite some time that the sea has plenty of treasures to offer, such as for the production of new medicines. But recently, increasing awareness that this is a huge potential source for human well-being offers yet another reason to care for our ocean ecosystems. Perhaps the potential financial benefits will serve as a better motivator than the much more significant indirect risks of an unhealthy ocean.

Oceans and Human Health is still in its infancy as a research field. Nevertheless, a growing number of recent studies are mapping the health benefits of coastal areas. We know, for instance, that people living near the coast are generally happier and healthier. We also know that this is not linked to more affluent coastal populations because the effect is even greater with socio-economically disadvantaged groups. They benefit even more from the coast than the rich do.

What causes this effect? That question is a lot more difficult to answer. Several hypotheses have been suggested, which we can divide into three large categories. The first (the 'Blue Gym') assumes the surroundings exert a psychological effect. The second focuses on exposure to bacteria and the resulting immunological benefits ('Old Friends'). The third and final hypothesis ('biogenic substances') centres on the effects of natural molecules excreted by marine organisms that enter the atmosphere through aerosols. Evidence has been found to support each of the three mechanisms, which leads us to suspect that the explanation may be a combination of all three.

Let's begin with the first hypothesis. Many international studies have attributed the positive effects of the sea primarily to psychological mechanisms: natural surroundings such as the coast and the sea are believed to reduce stress. The same has also been proven for other 'blue' and 'green' spaces, such as large lakes and forested areas. A British study from 2012 found indications that the positive psychological effect may be greater in socio-economically disadvantaged communities living on the coast than among those in green nature reserves. The study's authors explain this through a combination of less stress and more opportunities for walking, swimming, exercise, and other activities. They called it the Blue Gym hypothesis, comparing the sea to a kind of marine health club.

A study involving over 60,000 participants conducted along the Belgian coast by the Flanders Marine Institute (VLIZ), Ghent University, and the University of Exeter in the UK confirmed these results: anyone living less than five kilometres from the coast reports feeling healthier than people living further inland. The scientists studied several explanations. Perhaps the coast serves as a break from daily routine, encourages walking and exercise, creates a congenial atmosphere with plenty of social interaction, and offers improved air quality due to less traffic and industry. As it turned out, the first three explanations are insufficient. Coastal residents do not necessarily exercise more, do not have more social interaction, nor do they experience less routine-related stress than people further inland. The concentrations of particulates in the air were significantly lower than in the interior, but no clear relationship with health could be established. Perhaps the results were muddied by the complex fragmentation of natural environments in Flanders (not only dunes but also forests, heaths, and other nature reserves). So – although this may sound like an academic bromide – more research is needed.

The stakes are high, though: about 40% of the European population lives less than 50 kilometres from the sea. If the sea really does make us healthier and happier, then this is an excellent reason to appreciate and protect it.

If we can also prove that the sea promotes mental health, it would make our ocean even more relevant, as one in every six people in Europe experiences mental-health challenges. Recent research from the same academic groups suggests that nature promotes psychological recovery through changes in hormone regulation and the nervous system. While the exact mechanisms have not yet been discovered, everyone is familiar with the result: a walk in natural surroundings or along the shore does wonders for us, much more than in a busy city or an industrial setting.

Scientists from the VLIZ, Ghent University, and KU Leuven University conducted a study on how the mental well-being of people with access to coastal areas was affected by the COVID-19 pandemic and the resulting lockdown. They found a link between mental well-being and access to coastal areas, but not the frequency of actual visits to the shore. More specifically, coastal residents reported less boredom, less worry, and more happiness than inland residents. The study suggests that coastal proximity had a potential buffer effect on the adverse psychological effects of the COVID-19 pandemic, which supports the idea that the coast has a positive influence on well-being.

The Blue Gym was the first hypothesis explaining the sea's health benefits. However, several researchers claim that there are additional mechanisms that could improve our health, especially in the long term.

The second, immunological hypothesis builds on what the English researcher Graham Rook called the 'Old Friends' hypothesis. Rook has devoted his entire career to researching bacteria and other microscopic organisms that have accompanied humans throughout their evolution – our 'Old Friends'. These organisms are believed to be responsible for the proper functioning of our immune system. The theory is that our modern lifestyle is too hygienic, limiting our exposure to these old friends and thus making us more susceptible to allergies and auto-immune diseases. Rook's theory could also be extended to include our relationship with our old friends (microbial organisms) from the sea. Exposure to these organisms through water, air, or food could lead to positive effects.

The third hypothesis (exposure to beneficial biogenic substances through aerosols) is just as, if not more, interesting. It may initially seem similar to the 'Old Friends' hypothesis but focuses specifically on exposure to beneficial molecules produced by marine life. These substances can be ingested by people through food but also through the air, and it is the movement of the waves and the wind that causes these substances to enter the atmosphere as natural aerosols, what we call sea spray.

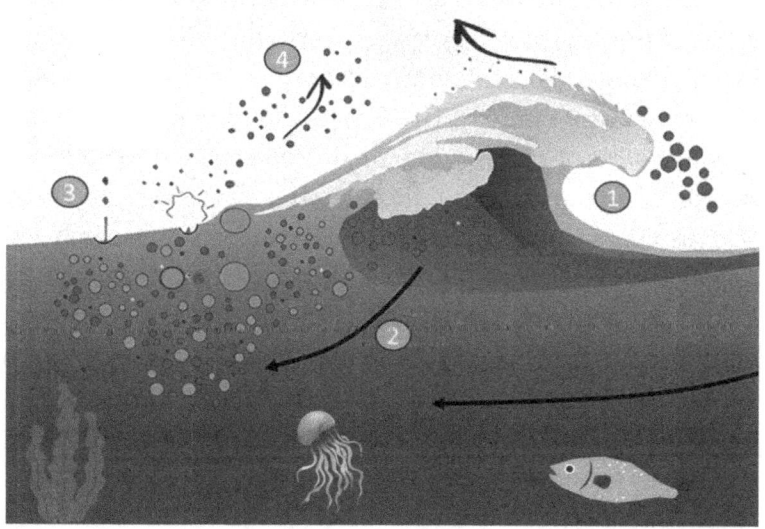

Figure 5. The sea contains countless species, such as bacteria, microalgae, seaweeds, fish, and large mammals, that produce bioactive molecules such as fatty acids and proteins. These molecules concentrate in the top layer of the ocean near the surface. When waves break (1), air bubbles form in the water (2) and rise to the surface. There, they take in the bioactive molecules and burst (3), transforming into aerosols or sea spray in the air (4). Figure based on Allen et al. (2020) Examination of the ocean as a source for atmospheric microplastics. PLoS ONE 15(5).

Researchers at Ghent University have tested the aerosol hypothesis. It was pioneering work, with lots of trial and error. They took air samples via air pumps that they carried on their bodies or installed at various locations along the coast. The samples were collected year-round in a variety of weather conditions and with different wind directions. The pumps collected the sea air, including all the molecules it contained, in filters. These filters were then transported to the laboratory, where the researchers tried to chemically identify the molecules in the sea spray. The researchers also cultivated various types of lung cells – including cancer cells – and exposed them to the sea-spray samples. The dosage was the equivalent of the amount of sea air you would inhale if you were to walk or exercise along our coastal regions for an hour every day for a month. One surprising result from the molecular research was that sea air could inhibit genes that are key to the development of cancer and high blood-cholesterol levels. This is an indication that the sea air could play a part in the sea's health impact on humans.

Before we call this a eureka moment, we should note that results from laboratory experiments cannot just be extrapolated to general effects on humans. We also still do not know *which* substances in the sea-spray mixture are responsible for these effects. However, this kind of research into the composition of sea air could pave the way to greater insight into the health benefits of the sea and the coast, which, in turn, could lead to new treatments.

Whatever the final verdict regarding the precise mechanisms, there is enough evidence to support the idea that these aerosols provide a possible explanation for the health benefits of sea air. In this context, a clarification is in order. Many people still think that ocean air is healthy because it contains lots of iodine. Although sea air does contain slightly more iodine than the air in inland and urban areas, this health benefit is a myth. The levels in sea air are simply too low to have any effect. Perhaps this myth originated during the rise of tourism and thalassotherapy along our coast, when scientists and doctors mainly studied the inorganic substances (such as salts) in sea air. As mentioned above, they established that iodine concentrations in the air are higher along the coast. They also knew that iodine is an essential element for human health, for example, regulating hormone production in the thyroid gland. An iodine deficiency can lead to a decline in cognitive functions as well as growth and development disorders. Thus, it did not take long to make a positive link with the sea. However, the daily intake of iodine required to avoid health issues is approximately 150 micrograms for an adult. Sea air contains only 5 to 30 *nanograms* per cubic metre of air, which amounts to about 0.5% of the daily recommended intake for the average adult – much too little to have any effect.

THE MEDICINE CHEST

For other medicinal substances, we must travel to a unique and fascinating ecosystem in our ocean. Anyone who has ever seen a nature documentary about the sea and has listened to master storyteller David Attenborough will undoubtedly be familiar with our colourful coral reefs. They are amongst the largest living structures on Earth: the Great Barrier Reef off the coast of Australia is even visible from space. Atolls such as the Maldives are mostly comprised of coral-reef skeletons.

Their most prominent residents and builders, coral polyps, are members of the Cnidaria phylum, which also includes jellyfish and sea anemones. Many coral polyps are reef builders, creating protective homes from calcium carbonate. These builder polyps filter water to capture plankton, their food source. The secret to many coral reefs is the polyps' partnership with zooxanthellae, algae that live under the coral polyps' 'skin' and give the corals their colour. They belong to the microalgae and carry out photosynthesis. This is the reason why we generally find coral reefs in shallow waters: just like other plants, the algae need sufficient sunlight for photosynthesis, through which they provide corals with essential nutrients, such as sugars. The corals, in turn, provide shelter and chemical compounds such as nitrates and phosphates that the zooxanthellae need for their metabolism.

Most corals live in colonies. All polyps in such a colony are genetically identical. When given enough room to build large reefs, the polyps form giant underwater cities. A large reef can house billions of polyps.

The capacity of coral reefs to support high biodiversity has led some to call them the 'rainforests of the sea'. They offer shelter and protection for countless fish, squids, crustaceans, algae, and turtles. They are a food source for birds, larger fish, and other predators. And for other species, the coral reefs function as a marine wellness centre, hospital, or apothecary. Coral reefs have actual 'cleaning stations' with cleaner fish and cleaner shrimp that eat pesky parasites off fish and turtles. This is a form of mutualism, a scientific word for a win-win situation. The customers have one less problem to worry about, and the cleaner fish score a meal. Even dolphins scrub their bodies against corals and sponges to keep their skin healthy.

The coral-reef system is also beneficial to humans. A coral reef is more than just a breathtaking biodiverse hotspot for snorkelling and spotting marine animals, or a fishing ground for coastal residents. Coral reefs also act as storm breakers that stop large waves. They are essential to the survival of many people in coastal areas. If they disappear, the beaches will become more vulnerable to heavy storm surges and tidal waves. Coral reefs are crucial natural coastal defence lines in the battle against climate change because they can grow to accommodate rising sea levels, just like other ecosystems such as mangrove forests and salt marshes.

Coral reefs are also useful as an apothecary. Many organisms that live in these ecosystems, such as sponges, sea squirts, anemones, snails, and the coral polyps themselves, use toxic substances to ward off enemies. Why do they need these poisonous substances? Almost every centimetre of a coral reef is inhabited; it is a complex city teeming with colourful residents and visitors. But that does not mean that it is always peaceful – the competition for food and space is enormous. Actual wars can break out, in which organisms fight each other to the death. Typically, if you found yourself in such a situation, you would pack your bags and move to another city or neighbourhood. However, many of the animals living on reefs, such as coral polyps, sponges, and sea squirts, cannot move. They swim about freely as larvae, carried by the currents, but as they reach adulthood, they find a place to settle down, attaching themselves to rocks or other hard surfaces. They become what we call sessile and stay rooted in place for the rest of their lives. This lifecycle is similar to that of the species we find in our local areas, such as mussels and acorn barnacles: free-floating as larvae, sessile as adults. It is tempting to draw a parallel with humans. A sedentary life has its advantages because you use less energy, but it also has drawbacks. You cannot flee from predators or intrusive neighbours that are after your food or living space. So what do you do? The solution that many coral animals have developed over the course of their evolution is to produce toxins. Or develop a disgusting taste or smell, which works too.

Sponges adopt this strategy too. They may look soft and harmless, but they are experts in biochemical warfare and can produce substances that kill viruses, fungi, bacteria, and even coral. The sponge *Siphonodictyon coralliphagum* drills through coral and produces a toxic substance that can kill polyps, freeing up more room to colonise.

Considering that the difference between poison and medicine is sometimes just a matter of dosage and application, we can infer that these coral dwellers can be used as inspiration to create beneficial medicine for humans. Sponges are particularly promising, and many of them have been extensively researched, but we are discovering increasing numbers of marine animals and plants that may produce potentially beneficial substances. The greater the variation in the substances they make, the greater the chance that some of them contain medicinal molecules.

How do we know which substances in the sea are beneficial to humans? In theory, that sounds as simple as making a cup of tea, but in practice, it is a lot harder. To start with, you collect the organisms you would like to study and which you suspect could be interesting. You then put them in a blender – unfortunately, we have not found a more humane method yet – and you extract the active ingredients by pouring a solvent such as ethanol over them. In reality, that analogy with making a cup of tea is not so far off: hot water also draws the chemicals from the tea leaves. You then analyse one part of the sample with chemical techniques and use another part to see how human cells react to the substance. Are there positive effects, or is the substance highly toxic? And if it *is* toxic, at what dosage? Ideally, researchers will eventually be able to produce a useful substance. DNA analyses have in recent years also been used to discover organisms that produce interesting bioactive molecules.

Meanwhile, a cancer drug has been developed based on sea squirts and a painkiller that uses a toxin from cone snails. Sponges are also useful: their antibodies are used in HIV- and cancer-treatment medicines. Many more applications will presumably be discovered in the future.

Drawing from nature is not a new way of developing medicines. Many modern drugs are based on medicinal plants and animals that have been known to us for centuries or even millennia. The earliest records of the use of natural products as medicines were found on clay tablets in cuneiform script from Mesopotamia (2600 BCE). The Ebers Papyrus (1550 BCE), an impressive Egyptian encyclopaedia, includes information about the medicinal properties of over 700 plants and their applications in gargling solutions, pills, infusions, and creams. The Chinese also described hundreds of recipes and medications in their book *Shennong Bencaojing* (1st century CE).

These, however, were usually applications of plants. Compared with terrestrial species, few marine organisms are currently used in medicine, but this field of research has grown tremendously in recent years. Some applications have been known for a long time. Irish moss (*Chrondrus crispus*) and false Irish moss (*Mastocarpus stellatus*), two species of red algae, have long been used in various remedies for colds, sore throats, and respiratory-tract infections such as tuberculosis. They were traditionally also boiled in water or milk for the treatment of kidney problems and burns. The extract of another type of red alga called laver (*Porphyra umbilicalis*) is effective against cancer if we can believe historical sources. This seaweed also grows in the North Sea.

These are all relatively old applications based on experience and tradition. Meanwhile, science has progressed a lot. This is also why it makes sense for us to look to marine life for potential medicine, because it has been around for 3.7 billion years and has had more time than terrestrial life to develop into the colourful diversity it presents today. As we saw in Chapter 3, the sea contains life forms (phyla) completely different from those found on land, which increases our chances of finding other, hitherto undiscovered molecules. The diversity in metabolic pathways is not only a product of their age and the time they have had to evolve, but also of their adaptation to extreme living conditions.

It is actually odd that, until about 1970, no one thought to study these many unknown species and find their potential uses. Maybe it had to do with our fear of the sea, or it could just have been ignorance. Whatever the case, this is quickly changing, in part because of the development of better analysis and sampling technologies.

The road to promising research into effective medicines is long. For now, only eight medicines based on marine substances have been clinically approved. Five of those eight approved drugs have been used for cancer treatment, while the other three are used to combat pain, the herpes virus (including cold sores, chicken pox, and shingles), and hypertriglyceridemia – better known as 'bad cholesterol'. For example, the drug Ziconotide is a peptide, the active ingredient of which was first discovered in a tropical species of cone snail. In December 2004, it was approved for the treatment of pain.

Worthy of note: seven of the eight approved medicines come from invertebrate marine animals. These creatures tend to produce highly bioactive substances, perhaps because they are so vulnerable and face stiff competition for food and space.

Many other substances still need to be fully tested. Plitidepsin, which is used for the treatment of cancer, contains a substance isolated from *Aplidium albicans*, a sea squirt native to the Mediterranean Sea. Another anticancer drug based on a sea squirt, this time *Ecteinascidia turbinata*, has already been approved. Others are no longer produced because they were not effective enough. The process of development is ongoing and, of course, involves a lot of trial and error.

SCALES AND SEAWEEDS

Invertebrates are not the only interesting things in the sea. Other marine organisms are used in a variety of applications, such as in cosmetics and the food industry. One problem, as you may already have guessed, is that many of the health-benefit claims are not based on solid scientific data. Another issue is that commercial exploitation entails risks that pose significant threats to the sea of the kind that we will describe in the following chapters. For instance, marine collagen is a substance used in food and cosmetics. Collagen is crucial for the functioning of our bodies, and a collagen deficiency can lead to ageing symptoms such as wrinkles, weak muscles, and painful joints. So, it comes as no surprise that collagen is increasingly used in products, especially now that a youthful look has become so important in our culture. It is big business. But what is the problem? Marine collagen is made from fish scales, and those fish need to be caught first. This means we need sustainable fishing (or farming) methods. More on that later.

There are also other applications in the food industry. One of the most well-known is omega-3 fatty-acid supplements. This too brings us back to the problem of overfishing. An interesting avenue that may circumvent the problem is using algae, both at a micro level (phytoplankton) and on a macro scale (different types of large seaweeds). For instance, there is already algae oil containing omega-3 fatty acids that offers a sustainable alternative to traditional fish-based supplements.

Algae are very popular today in many dishes and are rapidly becoming very popular in the culinary world. These are usually not microalgae (phytoplankton) but the larger algae that we will call 'seaweed' for convenience's sake. These seaweeds can be divided into three main groups: brown algae (Phaeophyta), red algae (Rhodophyta), and green algae (Chlorophyta). They owe their different appearances to their colour and photosynthetic pigments. Common examples of brown algae include bladderwrack or sea oak (*Fucus vesiculosus*), which we often find on our beaches, and sugar kelp (*Saccharina latissima*), which is used in macrobiotic cuisine and Asian dishes. The North Sea also contains red-algae varieties: the laver we referred to earlier that looks like a head of lettuce and grows on groynes, and Irish moss, which we now know is a traditional remedy for a variety of ailments. Of the green algae, we are particularly familiar with sea lettuce (*Ulva lactuca*) in the North Sea.

Most of these seaweeds are edible. In Japan, they have known this for centuries because many seaweed species are common in traditional cuisine. Perhaps you have heard of nori, kombu, wakame, ogonori, umibudo, or hijiki. These seaweeds are not only delicious but also incredibly healthy and nutritious. They include substances such as polysaccharides, proteins, polyphenols, carotenoids, phytosterols, and omega-3 fatty acids. Many studies show them to have a beneficial impact on certain metabolic illnesses, diabetes, cardiovascular diseases, cancer, and neurodegenerative diseases. Meanwhile, hundreds of other studies show seaweeds to offer a variety of health benefits, indicating that the undiscovered, hidden pantry of our marine apothecary is huge.

AS FIT AS A FISH

How healthy are fish and seafood for humans? Most food specialists generally consider fish and seafood to be an essential part of a healthy diet. Until now, the positive health benefits have mostly been attributed to fish oils, which are rich in omega-3 fatty acids. Its importance in preventing cardiovascular disease has been well documented. A meta-analysis of 20 studies involving hundreds of thousands of participants showed that 100 to 200 grams of fatty fish every week reduces the risk of death from cardiovascular disease by 36%. Fish such as salmon, herring, mackerel, and sardines are considered fatty fish.

Fish oil has other health benefits aside from a decreased risk of stroke: it lowers blood pressure, has anti-inflammatory properties, improves eyesight, and has positive effects on attention-deficit and hyperactivity disorders, schizophrenia, and dementia. Fish oil is also believed to help fight depression. Crustaceans and shellfish, by the way, are also a source of such useful oils.

But there is even more. Fish and seafood are important sources of micronutrients such as vitamin D and vitamin B12, as well as minerals, including selenium and iodine. We already know that vitamin D improves calcium absorption and bone formation, and can help prevent osteoporosis. Vitamin D may also help against psychological disorders, insulin resistance, and obesity, but these possibilities need to be researched further.

Fatty fish is healthy because of its oils, but for a healthy dose of iodine, you need leaner fish. Iodine is an important component of the thyroid hormones that play a fundamental role in regulating basic metabolism in humans. According to the World Health Organization, iodine deficiency is the most common cause of brain damage and mental impairment. Eating lean fish would help prevent these conditions. These include monkfish, whiting, and other members of the cod family, as well as flatfish such as sole and dab. Mind you: these substances are not miracle cures. The effects of diet are often complex and function in concert with a healthy lifestyle.

We have to issue yet another warning. Although algae, seafood, fish, and other treasures from the sea are naturally very healthy food sources, that may be changing in some respects. In introducing this topic, we are jumping ahead slightly to the next three chapters, where we discuss the effects of climate change, pollution, and overfishing. Many people wonder to what extent pollution negates the health benefits of fish and seafood, and it is a cause for concern. Warmer, more acidic, and oxygen-deprived oceans can lead to damaging algae blooms that produce toxins. Both the algae themselves and the toxins they produce can end up in aerosols, cancelling out the positive health benefits of sea air. They can also cause respiratory-tract conditions and acute breathing difficulties.

Poisonous substances from industry, such as methylmercury, PCBs, dioxins, brominated flame retardants (BFRs), perfluorinated compounds (PFCs), and pesticides accumulate in the food chain and can eventually end up in the tasty seafood and fish that we eat.

Sometimes, the concentrations of these toxic substances can reach levels that are catastrophic for humans. In 1956, several residents in Minamata, a fishing village in Japan, started having strange symptoms. They had problems with balance, weakened muscles, speech impediments, and other symptoms. Cats started to move erratically as if they were 'dancing', which is why people in Japan call it the 'dancing cat disease'. The cause was mercury-contaminated wastewater discharged from a chemical plant owned by the Chisso Corporation, which produced products such as acetaldehyde, a raw material for many other chemical compounds. A total of 900 people died, and about 2500 people suffered from severe chronic health problems. The actual numbers could be higher because not all of the cases were registered.

So, unfortunately, we cannot say that food from the sea is always healthy. Although its natural fatty acids, proteins, and minerals are good for us, it can also contain excessively high levels of toxic substances. Later on, we will also talk about plastics and microplastics, which are now ending up in our food chain. We do not yet know the extent of their impact, but their pervasiveness throughout our environment – on land, in the air and the water – is cause for concern among scientists.

In other words, we find ourselves walking a toxicological tightrope: ideally, we should be eating enough fish and seafood to enjoy their benefits, but not so much that we suffer from any negative effects. The problem is that, for many chemical substances and new pollutants such as micro- and nanoplastics, we do not yet know what the right balance is.

As science slowly progresses, climate change and the biodiversity crisis are shifting that balance as well. After our exploration of the ocean's beauty and riches, we now need to discuss the detrimental effect humans are having on this paradise. In contrast to what people once believed, out of a well-meant yet false idea of modesty, humans are significant enough to have an enormous impact on the world around them, and that includes our vast ocean. That negative impact will eventually come back to us like a giant boomerang: we can hurt the ocean, but the ocean is just as capable of hurting us. We reap what we sow. Fortunately, that same thought gives us reason for hope: if we have so much influence on the ocean, we can also use it in a positive way to turn the tide.

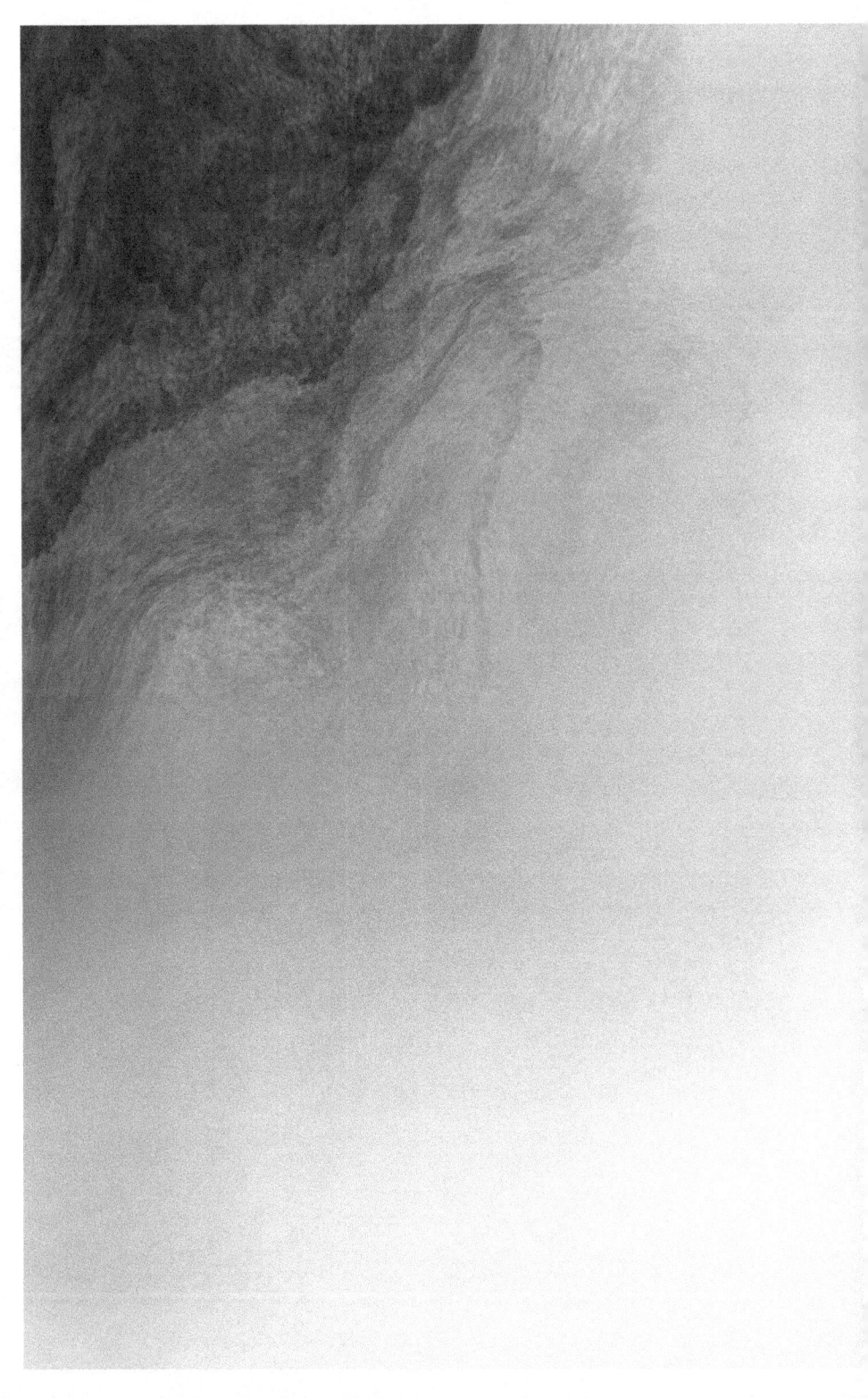

PART II
SWELL

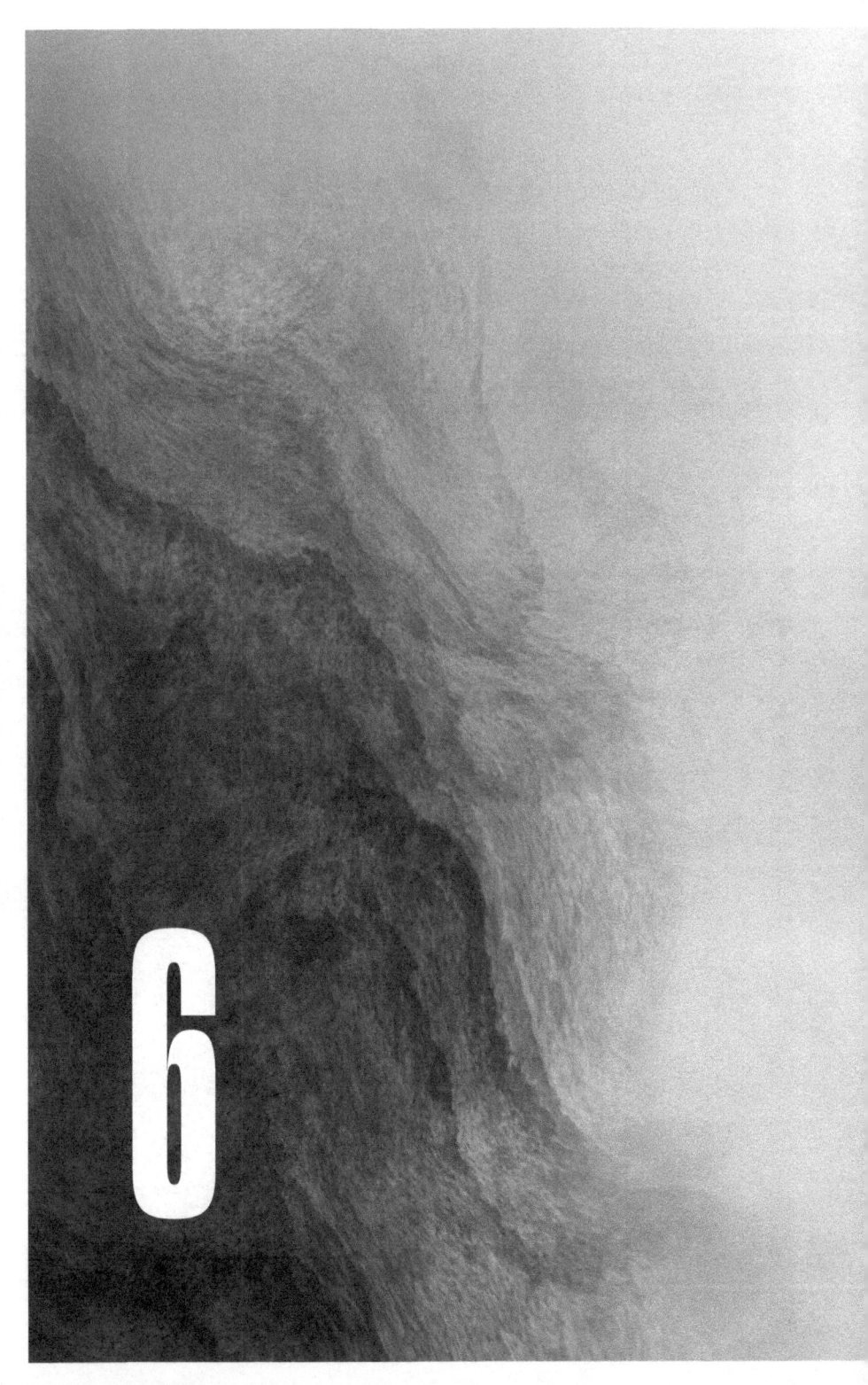

THE EARTH'S THERMOSTAT

> *"And so castles made of sand,
> melt into the sea, eventually."*
> **JIMI HENDRIX**

Time for a break. We arrive back on land and find ourselves sitting on the terrace of a café along the North Sea shore as a cool sea breeze provides relief from the sweltering summer heat. This is in strong contrast to the other side of the ocean, where hurricanes ravage the coasts of the Gulf of Mexico and the Caribbean every year. We also have violent storms, as every coastal resident is only too aware, and sometimes, the tail ends of these hurricanes reach the shores of Western Europe.

Hurricanes form at sea in the Atlantic Ocean but also in the north-western region of the Pacific Ocean, where they are referred to as typhoons. When they form in the Indian Ocean and the Southern Pacific, they are called cyclones. They draw their energy from warm surface waters.

As we have mentioned before, the air, the sea, and the land are inextricably linked. They constantly interact in a way that continues to amaze scientists. For thousands of years we have known that there is a link between the sea and storms or other violent weather phenomena. Remember Homer's *Odyssey*, in which the stubborn sea god Poseidon sends a storm Odysseus' way to ensure that the hero's ship is driven off course. However, only recently have we discovered the science behind it and the role that heat plays in this process. Moreover, we now need to consider a new factor, an uncomfortable truth that we cannot ignore: climate change.

As the sea heats up, hurricanes will become more powerful and make landfall in areas that historically have been spared. The latter is a potential issue for cities that are not prepared to 'receive' hurricanes reaching the shore. In Florida, where hurricanes are frequent, the infrastructure is adapted to accommodate such conditions, but this could be a problem in places like New York, Beijing, or Tokyo. One such example was Hurricane Sandy in 2012: the storm itself was only a category-1 hurricane, but it ranked fourth in history in terms of costs and damage because it made landfall further north. With warming waters, hurricanes will become stronger. What we do not know yet is whether the frequency of hurricanes will increase.

We need to gain further insight into the interaction between the ocean and climate change. We have already seen that the ocean provides major benefits as a marine apothecary, doctor, or therapist. But one of the sea's most critical benefits is one of which we have only recently become aware. For many years, little attention was given to this important quality, but in recent decades, we have come to realise how critical the ocean is in our fight against climate change. It stores heat and greenhouse gases, disperses them over its enormous volume, and exchanges these elements with the atmosphere.

Without its help, our future would look a lot less bright, and the warming of our atmosphere and the land would have started much earlier. Since modern pollution started in the Industrial Revolution of the 18th century, the ocean has stored no less than a quarter of all of the CO_2 produced by the combustion of fossil fuels and changes in land use. It has also absorbed an astonishing 90% of the extra heat produced by the greenhouse effect.

The silent sea can endure a lot, but it also has a breaking point. And the main culprits are what we call the 'deadly trio' for marine life: rising seawater temperatures, the associated decline in oxygen levels in the water (deoxygenation), and acidification. Without drastic action, the beauty and riches of our ocean, which we discussed in the previous chapters, are under threat. Less problematic for the ocean itself, but more so for the people that live on its shores, are rising sea levels, the resulting increasing salinity in agricultural soils in coastal areas, and the increase – both in frequency and intensity – of extreme weather phenomena. These are sea-related disasters, in other words, because the sea drives the weather that we experience on land.

Before we dive further into what climate change does to the sea, let's first discuss what the sea does to the climate. In a nutshell, the sea stores the excess heat that we produce. It is a buffer for fluctuations in temperature and an important ally in mitigating the effects of climate change and slowing down its undesirable consequences. The ocean is also an important carbon sink.

WHAT THE SEA DOES FOR THE CLIMATE

The sea not only regulates the weather but in the long term also the climate. It takes up about a quarter of all CO_2 produced by humans (and 35% of all fossil-fuel emissions) – initially through the solubility pump, which we came across in Chapter 1. It also stores heat: more energy from heat is stored in the top metre of seawater than in the entire atmosphere. In other words, it acts as a kind of air conditioner for the Earth, or more accurately, a thermostat, because it also tempers severe winters. It absorbs heat directly but also cools the Earth indirectly by storing the greenhouse gas CO_2. The ocean acts as a buffer for climate change, which is why it has taken so long for the effects of climate change to become apparent.

To understand how the ocean manages to do that, we need to go back to the carbon cycle. Carbon is the basic element of all life. You will find carbon compounds in the atmosphere, the ocean, and the soil, both in organic and inorganic forms. The amount of carbon in each area remains more or less in equilibrium despite – or perhaps because of – its constant exchange between air, water, and land.

One of the most important mechanisms for storing carbon is the biological carbon pump, which we already discussed in Chapter 1. Do not worry if you have already forgotten how it works: we will explain the entire process below. But this time, we are going to take things one step further and consider what happens when the pump starts to sputter. Phytoplankton is the protagonist, but certainly not the only player, in the biological pump. First of all, phytoplankton use photosynthesis to meet their energy needs for growth and reproduction, just like land plants. They store carbon and convert CO_2 into oxygen and other carbon compounds. For obvious reasons, phytoplankton can only be found in the upper layer of the ocean, in what we call the 'photic' zone. This layer receives enough sunlight for photosynthesis to happen.

Every evening, the largest and most spectacular mass migration on Earth takes place, although it is barely visible to the human eye. Billions of crustaceans, jellyfish, and other animal organisms migrate from the deeper layers of the ocean to the light-rich surface, spanning a distance of hundreds of metres. The Dutch scientist Jan Stel compares it to a huge Mexican wave travelling through the crowded ocean around the Earth as night falls. The zooplankton and the tiny fish eat the phytoplankton that spent the day storing energy through photosynthesis. As the sun comes up, they then migrate as one giant wave back down to the ocean's depths, leaving behind a trail of faeces, a variety of microorganisms, and straggling or dead phytoplankton that clump together to form 'marine snow' or 'sea dandruff', which in turn drifts down to the twilight zone. The result is a vertical conveyor belt in which the carbon first dissolves in the sea as CO_2, is then taken up by the phytoplankton, and ultimately travels to the ocean's depths through zooplankton. That is a crucial part of the biological pump.

At this stage, you may be wondering how zooplankton can travel upwards, given that we mentioned earlier that plankton, by definition, cannot move by themselves. We once thought that plankton were passive in their mobility, only moving with the water currents and the wind. However, we know better now. Even phytoplankton can move, either by releasing fats so they can change their buoyancy or even by swimming with their tiny tails. The surface has light for photosynthesis, while the depths have less light but more nutrients. We now even believe that half the phytoplankton in the sea regularly travels tens of metres to capture enough sunlight and nutrients. Those microscopic organisms take hours, days, and even weeks to do so. Some reproduce along the way so that their offspring can continue the journey.

We have already seen and discussed the famous marine snow earlier: a gigantic precipitation of dead organisms, plankton, protists, and faeces that clump into flakes and drift down to the deeper zones. This huge biomass ensures that each year, eight billion tonnes of carbon migrates from the atmosphere into the deep sea. Bear in mind that in addition to this flux, we also have the solubility pump and the carbonate pump, which not only store carbon but also release it back into the atmosphere (see Figure 6).

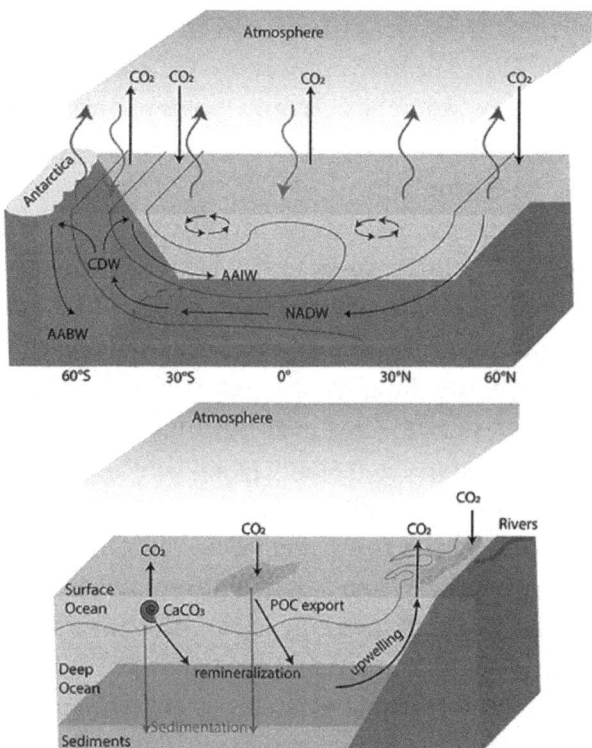

Figure 6. The carbon pump in the ocean. The top image shows the solubility pump, which takes CO_2 from the atmosphere and transports it through currents from the surface to the deeper zones – and back again. The bottom image represents complex biological pumps. CO_2 is absorbed at the surface by phytoplankton and sinks as particulate organic carbon (POC) into the deep sea. On the other hand, calcium carbonate produced at the surface releases CO_2. Organic carbon is remineralised in the deep sea to CO_2 and brought back to the surface via upward currents. Source: Landschützer (2023).

The net result of all these processes is that the ocean takes up 2.5 billion tonnes of carbon each year. That is about two to three times as much as the annual emissions of all aircraft worldwide. Without these processes, CO_2 concentrations in the atmosphere would be 200 ppm (parts per million) higher than they are today, and temperatures on Earth would already be 4 to 6 degrees higher.

Larger marine animals also contribute to carbon transport. Each dead animal that sinks to the bottom of the sea transports carbon. We have already seen how dead whales take massive quantities of carbon with them to the bottom of the ocean. But the role of jellyfish – referred to by scientists as 'gelatinous plankton' – has been seriously underestimated. Researchers estimate that at any given time, jellyfish species store up to about 510 million tonnes of carbon, a stock that can produce between three and seven billion tonnes of carbon waste annually. When jellyfish die en masse in so-called jelly-falls, an immense amount of carbon ends up on the sea floor – about two billion tonnes per year.

Another underestimated factor is the role of seaweed, such as *Sargassum* in the great Atlantic Sargassum belt. Fertilisers from the food industry flow through the Amazon, Congo, and Mississippi rivers into the Atlantic Ocean, where they feed the *Sargassum*. These flourish and form giant seaweed fields that are a plague for many beachgoers in Mexico and the Caribbean Sea. As they wash up on shore and decompose, they release a foul rotten-egg smell. In the shallower coastal areas, they can also cause oxygen depletion resulting in fish deaths. This uncontrolled seaweed growth can also kill corals and seagrass by blocking out sunlight. More relevant to our story, though, is the fact that these seaweeds contain lots of carbon. Scientists are currently researching the possibility of pushing them down to the deep-sea floor using robots so that we can remove the carbon from circulation for at least a couple of centuries.

We have not yet explained the last step in the process. What happens to all those carbon-rich particles from plankton, plants, and animals that end up on the seabed? When organisms meet their inevitable demise and do not end up in the stomachs of other animals or broken down by bacteria, the carbon they are made of sinks to deeper reservoirs called carbon sinks. There, they lie buried for millions of years, eventually turning into oil or natural gas through a range of complex processes. We are now witnessing what happens when we access and burn those ancient fossil-fuel stores of oil and natural gas: a massive release of CO_2 and other greenhouse gases that absorb the sun's warmth, heating the Earth in the process. In other words: climate change.

The biological pump is a fantastic process. It is a carbon sink that deserves our full consideration or that we should at least not disrupt. Without these physical-chemical processes in the ocean and the biological pump, CO_2 levels would be 50% higher than what they are today. The question now is whether the pump is slowly starting to sputter because of all the pressures it is currently under. The healthier our oceanic ecosystem, the richer the biodiversity, the better the biological pump works, the more CO_2 the ocean can draw from the air, and the more favourable the circumstances for humans on Earth. This, once again, proves that everything is connected: sea, air, and land. The downside to this interconnectedness is that a weak link, such as a warmer or more acidic sea, can jeopardise the entire system.

What does climate change mean for the ocean? The answer is: nothing good, because the worldwide scientific results summarised in the Intergovernmental Panel on Climate Change (IPCC) reports about the ocean are disconcerting, to say the least. The changes that we see taking place in the sea today are unprecedented. We are specifically referring to the 'deadly trio': warming, acidification, a decrease in oxygen levels, and the interaction between these three. All three elements also create problems for the biological pump.

THE WARMING OF THE OCEAN

As the Earth heats up, so does the ocean. Each year, more heat is added to and stored in the ocean. This is not new: ever since the 1950s, humans have been recording a rise in temperature unprecedented in human history, with new records being set in the last ten years. The long-term trend indicates that global warming is accelerating. Additionally, this warming takes place everywhere, from the surface of the ocean down to the deep sea. The sea absorbs the heat at the surface and distributes it over the various depth layers through currents, mixing, and convection (heat transfer caused by changes in temperature or pressure, like a radiator).

This not only has consequences for the currents, the chemistry, and the biology of the sea itself, but also for the weather on land. A warmer sea leads to more extreme events, such as more extreme rainfall accompanied by flooding. But equally common are droughts and large forest fires. In recent years, the media headlines regularly report extreme weather conditions all over the world. Storms and hurricanes are becoming more powerful and more frequent. There are always variations, and it is always difficult to attribute specific events to climate change, but it does seem to be a long-term trend, and the peaks are getting more extreme. It is like when you are walking the dog: the dog veers to the left and right to sniff the grass, but eventually, it will continue along the path. Similarly, there are constant variations in temperature, but climate change determines the course in the long run.

We may also be faced with potential unforeseen domino effects. One of these is what we call albedo, the ability of a surface to reflect light. As the ice caps melt through global warming, less sunlight is reflected back into the atmosphere, accelerating the heating process. Global warming is melting the ice both on land and at sea. In particular, the melting of the polar caps on land (Antarctica and Greenland) can lead to rising sea levels. Perhaps you have heard that the melting of North Pole ice (floating sea ice) is not responsible for rising sea levels. But that is not entirely accurate. Unlike land ice, sea ice will not directly cause a sea level rise, just like the water level will not change in a glass filled with ice cubes as they melt – a phenomenon known as Archimedes' principle. But when the ice melts, the albedo effect is reduced, accelerating the warming process. Moreover, warmer water expands and takes up more volume. This is called 'thermic expansion', something we will come back to later.

Another potential unforeseen consequence concerns the ocean's conveyor belt which we mentioned earlier, the thermohaline circulation. This giant Atlantic sea current transports almost twenty million cubic metres of seawater per second, a flow almost a hundred times the capacity of the Amazon River. Warm surface water flows north and returns south via a deep, cold underwater current. The heat transfer involved is enormous: more than a million gigawatts, which amounts to about a hundred times the energy that we humans consume annually. That heat is returned to the atmosphere on the European end of the northern Atlantic Ocean, which has a lasting effect on our mild climate.

In 1987, the American geochemist Wallace Smith Broecker, who developed the concept of the ocean conveyor belt, published a controversial article in the scientific journal *Nature* about 'unpleasant surprises in the greenhouse'. Even Hollywood picked up on the concept in the rather mediocre film *The Day After Tomorrow* from 2004. The hypothesis is that the large influx of cold, fresh water from the polar caps can slow down or even stop the conveyor belt, with catastrophic consequences for our 'mild' sea climate. Today, we know that the Gulf Stream cannot stop circulating and that we are not faced with an immediate ice age. However, at the moment it is too early to say what will happen in the long term. We are still undecided on whether we have actually measured a decrease in the flow rate. We have seen some disconcerting indicators from measurements of the surface water, though: the North Atlantic region is the only region that has cooled down since the 19th century, while at the same time, we see extreme heating off the North American coast.

In its report on the ocean, the IPCC concluded that the circulation has weakened since 1850–1900. There are also indirect indications that the Gulf Stream is slowing down. The Florida Stream at the beginning of the Gulf Stream is weaker than it used to be. The waters of the South Atlantic Ocean have become saltier due to increased evaporation. Scientists predict that the circulation may weaken up to 40% by 2100 and potentially become unstable. These may be the consequences of rising temperatures in the ocean, but we do not know for sure.

Perhaps you may be wondering: how on Earth do you measure how warm the ocean is? If you take a bath and add a bit of warm water after a while, you can sense the temperature difference between the bottom and the top layers in a relatively small volume of water. So, how can you know what the temperature is in such a gigantic, multi-layered body of water like the ocean? The answer is the Global Ocean Observing System (GOOS), an international programme where countries collaborate to gather marine observations using standardised methods.

This technology has evolved rapidly. In the early 20th century, scientists still had to use Nansen bottles at great depths to collect water samples. A Nansen bottle was a tube, made initially from brass, which was attached to a cable and weighted down with a heavy weight. The bottle contained a thermometer that fixed the mercury at the depth of the sample so the temperature at that depth could be read. Starting in 1940, the Nansen bottles were replaced by mechanical bathythermographs, instruments with a temperature sensor that linked the measurement to the depth. The bathythermograph was developed by Carl-Gustaf Rossby, who entrusted its further development to his student Athelstan Spilhaus (the same Spilhaus who created the Spilhaus Projection; see Figure 1 in Chapter 1).

The revolution came in 1999 with the launch of Argo. Argo is an international programme that collects data about temperature, salt content, and currents through a system of robotic data-collection probes called floats. The probes drift on the currents and take measurements at depths of up to 2,000 metres.

About 4,000 floats are currently active. Ocean scientists and engineers are working hard to develop improved versions that will take measurements at greater depths and collect biogeochemical data (where biology, geology, and chemistry meet).

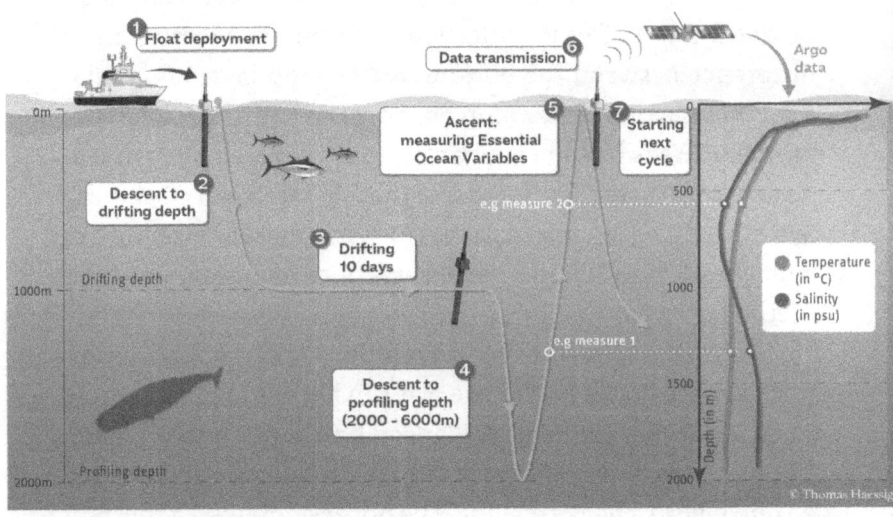

Figure 7. The Argo programme uses over 4,000 floats worldwide to collect information about the ocean's temperature and salt content. The project is one of the largest data sources for climate-change research. Source: VLIZ, *De Grote Rede 53* (2021), from the Argo programme, © Thomas Haessig.

The different measurements over time clearly show that the ocean has got warmer in every zone throughout the water column. From 1990 onwards, we see that this warming trend is accelerating. When comparing 1850–1900 with 2010–2019, we can see that our Earth's atmosphere has warmed up by 1.07°C. During that same period, the temperature of the ocean has risen by 0.7°C. We have already mentioned that the ocean absorbs 90% of the excess heat we produce (through the greenhouse effect), with the rest being trapped in the land (5%), ice or the cryosphere (3%), and the atmosphere (2%). That 90% is currently mostly found in the upper surface layers but some of it is also in deeper zones. The prediction for 2100 is that the atmosphere's temperature will continue to rise by about 1.5 to 4.3% compared to the period 1850–1900, depending on which scenario we adopt and the amount of greenhouse-gas emissions we continue to produce. The world's ocean follows this pattern, as does the North Sea. Specifically for our local region along the shores of Western Europe, we can expect a continued rise in seawater temperatures of about 0.05°C per year, or 1°C every 20 years.

We need to be aware that the real effects of climate change will only become apparent after a while. With its large volume and the water's high heat capacity, the ocean reacts slowly to temperature changes in the atmosphere. The scientific term for this is thermal inertia. It is exactly because the ocean reacts slowly to historical and current carbon emissions that the acceleration and expansion of warming we are seeing in its depths give us reason for concern.

In a realistic emissions scenario, climate scientists predict that the heat stored in the top two kilometres of the ocean in the period 2017–2100 will be five to seven times that of the 1970–2017 period. This is not an immediate cause for despair, but the irreversible nature of the warming of the ocean is a fact that we need to consider.

Note that the above figures are averages. There are also extremes: marine heat waves. Extreme heat stress is a problem for marine organisms, particularly sessile (attached) species that cannot escape. This includes sea squirts, coral polyps, mussels, and macroscopic algae: they spend their whole lives tethered to their habitat and must endure all these changes. These heat waves are more common and extreme in the tropical and arctic regions, but we are also starting to see some of their effects in 'temperate' parts of the world. The Mediterranean Sea is a much-discussed region in this respect, but the North Sea and other semi-landlocked seas in Europe will not escape these effects. Studies conducted by Ghent University and the Flanders Marine Institute have recently shown that, during the hottest summers in recent years, the populations of the most critical zooplankton species in the Belgian part of the North Sea collapsed.

The heat waves are mostly attributable to human activity. Eight of the ten most extreme marine heat waves occurred after the year 2010. And in European seas, they have become twice as common since 1982. They are also more intense and long-lasting, and researchers expect that, as we approach 2100, they will become 20 to 50 times more common worldwide than in the 1850–1900 period. Heat waves that once appeared once every hundred or thousand years will now occur every ten or hundred years as the global temperature rises by 1.5°C, and once every year or every ten years once the temperature increases by 3°C.

Although we just said that currents disperse the heat, there are still large regional differences in ocean warming. The Pacific Ocean generally sticks to the global average. The warming effect is strongest in the tropical and temperate zones of the Atlantic Ocean and the Southern Ocean. The Indian Ocean has a less marked increase in temperature and was even relatively mild until the late 1990s. At the same time, the sea ice at the North Pole is melting quite rapidly, and that alone has an indirect effect on sea levels, as we saw earlier with the albedo effect and thermal expansion. Sea-ice levels in the northernmost regions reached their lowest point in the last thousand years in September 2020. Whatever scenario we adopt, an ice-free North Pole at the end of the summer period will occur at least once before 2050. Another problem is that the warmer water from the Atlantic Ocean is also causing the ice cap on Greenland to melt. That is land ice that *will* directly cause a rise in sea levels, and its melting will significantly disrupt existing climate patterns.

Annual or ten-yearly variations in surface temperatures and seawater are related to changes in the regional or local climate. We know that high water temperatures in the North Atlantic Ocean are linked to increased hurricane activity, the summer climate in Europe and North America, and the monsoons in the African Sahel, India, and Brazil. The warming of the ocean can disturb ocean currents, changing the distribution of excess heat and causing significant changes in the local climate.

THE CHRIST CHILD AND THE LITTLE GIRL

As the seawater warms up, existing variations in weather patterns grow more extreme. An important mechanism involved is El Niño: a rapid rise in temperature in the Pacific Ocean off the coasts of Peru and Ecuador. This rise in temperature occurs once every few years and has global consequences. As the surface temperature heats up, less cold water wells up from the depths, leading to a decreased supply of nutrients and consequently fewer plankton to feed the fish.

This phenomenon is not new. Anchovy fishermen from Peru call this period El Niño ('the little boy' or 'the Christ child') because this phenomenon often occurs around Christmas. During La Niña ('the little girl'), the opposite of El Niño, the surface water cools down in the central and eastern tropical Pacific Ocean. This results in more fish in the region and – again – visible consequences around the globe. El Niño and La Niña are opposites, dance partners in the meteorological tango that goes by the scientific name of the El Niño Southern Oscillation (ENSO).

As with other weather phenomena, climate change will also lead to more frequent extreme El Niño events. The ocean has to blow off steam somehow. If the water is warmer than elsewhere, all that stored energy needs to go somewhere. The media headlines during the super El Niño in 2015 were as plain as day: 'Nearly One Million African Children Are Malnourished Thanks to El Niño', 'Wildfires Burn Up Western North America', and 'Godzilla El Niño Wreaks Havoc: South America Floods As Australian Forests Burn'. These headlines are a testament to the fact that El Niño causes chain reactions that influence weather patterns across the globe. In Europe, the effects of El Niño were modest in comparison. There were little or no visible effects in our region. However, a study published by the Royal Netherlands Meteorological Institute did indicate an indirect relationship between El Niño and the heavy spring rainfall in the Netherlands and Belgium. Since 1856, almost every El Niño period has resulted in a wet spring in Western and Central Europe.

The effects are less dramatic with La Niña, El Niño's cooler counterpart. Still, this phenomenon is linked to periods of drought in the United States and an increase in the development of hurricanes in the Caribbean and the Atlantic region.

El Niño and La Niña are intensifying with climate change, as are their effects on sea and land. That is because both are triggered by increasingly warm conditions in the ocean.

El Niño is not the only phenomenon of its kind. There are other similar see-saw patterns, or oscillations, that influence the Earth's weather. We also have the Arctic Oscillation (AO) and the North Atlantic Oscillation (NAO). They may be small change compared to the ENSO, but they do play a significant role in the weather patterns in our region. The NAO is driven by a high-pressure area above the Azores, the Azores High, and a low-pressure area that usually lies above Iceland. Oscillations in the strength of these two fields determine the force and direction of the winds above the North Atlantic Ocean.

The NAO also partly determines the location and intensity of what we know as the jet stream above the North Atlantic Ocean. There are generally two states: a positive and a negative state. During a positive NAO, the difference between the sub-polar low above Iceland and the tropical high above the Azores is great. That leads to a strong westerly wind, which generally leads to more precipitation and milder winters in the Low Countries. During the negative phase, the difference is smaller, weakening the westerly airflow. The result? Dry periods in Europe with cold winters.

As with El Niño, the strength of the NAO is expressed using an index, the NAO index. A key difference is that the NAO is primarily an atmospheric phenomenon (although it does have some limited effects at sea level), while El Niño influences both atmospheric and oceanic circulation. The NAO index varies greatly because the index can jump from a positive to a negative value within the space of a month. On the other hand, periods where the positive or negative mode dominates can span several years. This variation can then be linked to periods with warmer or colder winters in our region. Researchers have also discovered that El Niño and NAO team up. For instance, the winter of 2009-2010 showed significant deviations in snowfall in the Northern Hemisphere. Experts concluded that this deviation was the result of an interaction between El Niño and NAO. El Niño that year caused more precipitation in the southern part of North America, while the negative NAO phase led to a cold winter in the eastern part of the country. The result was extreme snowfall on the east coast of North America. In contrast, other regions experienced less rainfall or slightly higher temperatures than usual.

There are also oscillations and jet streams elsewhere in the ocean, for instance in the Southern Ocean around Antarctica. They are essential to the thaw and regrowth of the ice caps in Antarctica. The study of those oscillations and their interaction with the ENSO, for instance, is critical for understanding what will happen to land ice and, in turn, sea levels in the future.

BLEACHED CORALS AND FLEEING FISH

Anyone who has ever seen bleached coral will know how detrimental rising ocean temperatures are to marine ecosystems. Since 1995, we have lost half of Australia's Great Barrier Reef. Coral bleaching is the result of the loss of the algae (the one-celled zooxanthellae) that live in symbiosis with the polyps. When the algae experience heat stress, they secrete toxins. To protect themselves from these toxins, the polyps show the zooxanthellae the door, but then they also lose those symbiotic benefits. As the evicted zooxanthellae algae are responsible for the reefs' colour, the coral structures turn pale. After a while, the corals die, and the abandoned reefs turn into ghost towns.

We see marine life under pressure everywhere in the ocean. Some cold-water species such as cod must travel towards the poles to survive. That has enormous consequences for the fishing industry. The negative impact is most severe in the least developed economies, where the dependency on small-scale local fishing as a source of food for coastal populations is greatest. At the same time, these populations have fewer means to adapt to changing conditions.

Changes in habitat, shifts in the distribution of species, and loss of biodiversity are well underway. One example is the loss of tropical habitat through heat stress – the more complex the organism, the lower the heat tolerance. In the warmer parts of the ocean, the 'higher' or more complex organisms will become extinct first. The factor with the greatest potential to cause extinctions is the interaction between rising temperatures and oxygen depletion.

The interaction between ocean warming and pollution also deserves extra attention. It can cause hazardous algae blooms, a phenomenon that has become more frequent and widespread since the 1980s, with disastrous consequences for food safety, tourism, and human health.

THE SUFFOCATING OCEAN

In addition to rising temperatures, decreasing oxygen levels is another consequence of climate change. This phenomenon deserves special mention and an explanation.

As the ocean's water temperature increases, the density of the surface water decreases because warm water expands. Something similar occurs when ice melts because this adds fresh water, which is less dense than salt water, to the mix. The result? Water layers with different densities do not mix as well. Scientists use the term 'stratification': the formation of distinctive barriers between water layers. Warmer water containing less salt is lighter than cold salty water. Oceanic layers of different densities form on top of each other, impeding circulation across those layers. The greater the density difference, the less exchange (of oxygen and nutrients) between the surface waters and the deeper ocean. So, differences in salt content, just like temperature, change the density of seawater and the currents.

With this poorer circulation, the oxygen-rich water is unable to reach the deeper layers of the ocean, which can significantly impact the ocean's ecosystems and the fishing industry. The influx of nutrients from the deep sea also decreases, which means there is less food available at the surface for the phytoplankton, which in turn has a significant influence on the foundation of the food chain and, ultimately, on fish populations and fishing. Since 1960, there has been an estimated decrease of about 5% in what we call primary production (the production of organic material through photosynthesis).

The decreased circulation and decline in phytoplankton are causing the ocean to lose oxygen, which is already happening as the ocean heats up because oxygen is less soluble in warmer water. Between 1970 and 2010, the global decrease in oxygen levels in the top 1,000 metres of the ocean amounted to a few per cent (0.5 to 3.3%). The volume of what we call oxygen-deficient zones has also increased by 3 to 8%, in part due to warming climates. And, as is the case with rising sea levels, the decreased oxygen levels in the ocean can last thousands of years, even if we do change course now.

Oxygen is not a big problem for land animals such as humans; we will always have enough oxygen to breathe. Sadly, we cannot say the same for life in the sea. As the water gets warmer, it will contain less oxygen. For marine life, hypoxia (too little dissolved oxygen) leads to stress, emigration, and death. That is particularly true when it comes to heat waves in shallow or (partly) landlocked bodies of water. However, a lack of oxygen has also been demonstrated at depths of 1,000 metres.

It is estimated that the loss of oxygen, fewer nutrients, and decreased carbon removal will lead to a 15% decrease in the world's biomass by the end of this century compared to the 1986–2005 period. The consequence is a decrease of 20–25% in the maximum possible fishing catch, a trend that has already been set in motion and currently stands at 4% (1930–2010). In the even longer term leading up to 2300, the decline in fish populations in North Atlantic waters could even be as high as 60%.

ACIDIFICATION

Ocean acidification is referred to as 'the other CO_2 problem' besides the greenhouse effect. While this phenomenon is relatively unknown to the public at large, it is definitely a source of concern for climate and marine scientists. And although even those scientists admittedly do not know much about the issue, the broad outlines are starting to take shape.

Acidification as a climate problem is unique to the sea and has little to do with the rising temperatures caused by the greenhouse effect. For its root causes, we need to look to chemistry. Do not panic; we will try to keep things as simple as possible. Much of the excess CO_2 that we produce through the combustion of fossil fuels (around a quarter, as we saw earlier) ends up in the seawater, initially via the solubility pump. When CO_2 dissolves in water and reacts with it, it forms carbonic acid (dihydrogen carbonate), which then dissociates into bicarbonate ions and hydrogen ions (H+). The result: the water becomes more acidic. Not so much that the water suddenly turns into vinegar. Seawater is still alkaline because its pH value is still above 7 (we only refer to anything as acidic when its pH value drops below 7).

Since the late 1980s, the ocean's pH has generally dropped by 0.017–0.027 units per decade. Between 1751 and 2021, the sea has seen an increase in hydrogen-ion concentrations of 30%, corresponding to a decrease in pH value from 8.25 to 8.14, the lowest in 26,000 years. A decrease of just 0.11 units may not seem like much. However, pH is like the Richter scale for measuring the magnitude of earthquakes; it is a logarithmic scale and not a linear scale. Therefore, a small change can make a big difference. A decrease of just one unit of pH represents a tenfold increase in the number of hydrogen ions in the ocean. If we follow the business-as-usual scenario (keeping emission levels the same), the pH will decrease to about 7.8 by the year 2100. These differences may sound small, but seldom in the past two million years have such values been seen.

Is it bad that the ocean is acidifying? Yes, especially for all calcium-based marine life. Anyone familiar with domestic chores will know that acid and calcium do not go well together. That is why we use vinegar or some other type of acid to descale our coffee makers. It is estimated that about half of all marine life forms, particularly organisms that have calcium (calcium carbonate) in their skeletons or that produce calcium plates or shells, will suffer because of acidification. In a more acidic ocean, calcium is less readily available.

In the ocean, species such as shellfish, corals, sea snails, sea urchins, all kinds of plankton such as the coccolithophores and foraminifera, but also crustaceans and fish have skeletons made from calcium carbonate or calcium. As the water grows more acidic, these organisms will have more trouble forming their skeletons or shells.

On top of all that, we are not certain whether the carbon pumps will continue to function in a sea that is increasingly warm and acidic. You would think that more CO_2 is a good thing, considering that it is required to fuel the food chain. However, the equilibrium between the biological pump and the carbonate pump is highly complex, and there is simply too much CO_2, which means that not all of it can be absorbed by the phytoplankton: the excess carbon accelerates the acidification process. And it gets even more complicated: in a warmer ocean, less CO_2 would be dissolved, which would slow down the acidification process. So, there are a lot of opposing forces at play that we have not even begun to fathom, which makes it all very confusing.

Scientists consider sea butterflies and other winged pelagic snails to be good indicators of ocean acidification. When we introduced these creatures into seawater with the pH value predicted in 2100, their shells dissolved within 45 days. Reason enough to make sure this does not happen.

A TSUNAMI IN SLOW MOTION

Anyone who has seen the footage of the devastating tsunami in Indonesia in 2004 will have these images engraved into their memory. On Boxing Day, a massive earthquake with many aftershocks shook the seabed of the Indian Ocean. What followed was a tsunami that destroyed almost everything in its path. In 2011, a similar tsunami occurred after an earthquake in Japan. Treacherously, the sea first recedes, as if everything is fine, to then roll ashore with a devastating crash. Moments like these confront us with the primaeval, all-consuming power of the sea and our powerlessness as mere humans to stop it. While tsunamis are not always the massive monster waves you see in disaster films like *The Day After Tomorrow*, they still cause enormous damage because they can reach so far inland. The 2004 tsunami claimed more than 200,000 lives. The 2011 tidal wave in Japan totalled 20,000 victims and resulted in the Fukushima nuclear accident.

Such natural disasters are unavoidable. The same cannot be said for the gradual but just as treacherous sea-level rise caused by climate change, which is currently well underway. A sea-level rise of 20 centimetres has already been recorded, and the water level is rising faster than it has in the past 3,000 years. Official estimates from the IPCC assume a rise in sea level measuring between 60 centimetres to a metre by the end of this century if we do not take drastic and immediate action, but other predictions assume that sea levels will have risen by several metres by then. Especially if land ice starts to melt or break off into the ocean on a mass scale, those levels could rise very quickly. And when you add to that the fact that 40% of the world's population lives on the coast, this is cause for concern.

Think back to the submerged Doggerland and Testerep that serve as a warning today.

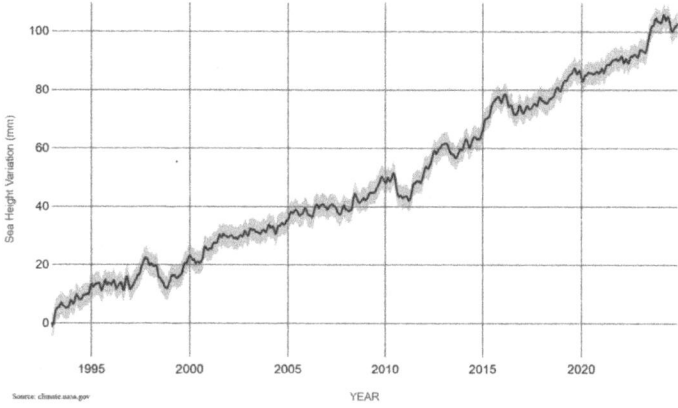

Figure 8. Global average sea level has risen more than 20 centimetres since 1880. The rate of global sea level rise is accelerating: it has more than doubled from 1.4 millimetres per year throughout most of the 20th century to 3.6 millimetres per year from 2006 to 2015. Since 1993, global sea level has increased by 10 centimetres, Source: Satellite sea level observations, NASA's Goddard Space Flight Center

It is not so much the averages that are worrisome, because countries like the Netherlands and Belgium show that the sea can be kept in check with broader beaches, higher dikes, and proper water management. What does concern us are the peaks, when the sea is already elevated and weather conditions get more intense. Storm surges such as the one that flooded Zeeland in the Netherlands in 1953 could become more frequent, and tsunamis such as those in Indonesia and Japan could roll further inland, not to mention the many poor low-lying countries that have trouble defending themselves against the rising sea, such as Bangladesh and the small island states in the Pacific Ocean, where this problem is beginning to emerge.

What causes the level of the sea to rise? You would think that it is mainly caused by melting ice, but for now, the biggest cause – and this may surprise you – is the simple fact that water expands as it heats up. This factor explains up to 50% of rising sea levels so far. Of the remaining 50%, 22% comes from meltwater from shrinking glaciers, 20% from the melting ice caps in Greenland and Antarctica, and 8% from groundwater extraction. But those proportions are changing, and we expect the ice caps to eventually play the biggest role. The remaining land ice is still good for a potential sea-level rise of about 70 (!) metres, particularly in Antarctica (58 metres) and Greenland (8 metres).

Just like a giant tanker takes a while to change course once you turn its rudder, the many effects of climate change will only become apparent over time. Climate models predict that sea levels will continue to rise for centuries, millennia even, as warmth continues to be absorbed by the sea and we continue to lose our ice caps. In 2300, those predictions assume an increase of 1 extra metre with low emissions and between 2.3 and 5.4 metres with relatively high emissions. As for the even longer term, let's say in the next 2,000 years, sea levels are estimated to rise up to 22 metres, depending on how much the Earth continues to heat up.

KICKING THE FOSSIL-FUEL HABIT

All this data should be enough to convince us that we cannot let things get that far and that we should do something as quickly as possible. The 10% of the world population that currently lives less than 10 metres above sea level and particularly the 65 million residents of small low-lying islands would be eternally grateful if we did. There are certainly enough reasons to swing into action. The IPCC does not rule out the possibility of the South Pole ice caps accelerating into a state of irreversible instability. This phenomenon, recorded in Amundsen Bay (West Antarctica) and on Wilkes Island (East Antarctica), could disrupt sea levels even more: 2 metres by 2100 and about 5 metres by 2150.

Moreover, rising sea levels bring a whole host of other problems and effects. Highly relevant for the Low Countries, and particularly our coastal polders or plains, is salinisation. Rising sea levels increase the pressure of seawater on the land and push salt water further inland through our dunes and polders. Layers of groundwater used for agricultural lands and drinking water become salty and, therefore, unusable. When there is a severe drought and there is little river water flowing into the sea, seawater can flow upriver even further inland.

These are all very good reasons to tackle the problem now. But how can we do that? As mentioned before, the solution is actually simple: first, stop burning fossil fuels, and change our land use and dietary habits. A healthy ocean can be part of the solution here. Think of its contribution as a potential energy source, carbon storage facility, and natural coastal defence line. To keep the ocean healthy, or perhaps even improve its health, the best option seems to be the recovery of marine biodiversity and the sustainable rebuilding of biomass. Only by combining forces, setting strong policies, and evoking a sense of urgency can we keep the ocean as a powerful ally.

To do that, we especially need to kick our fossil-fuel habit. Otherwise, the ocean will no longer be able to fulfil its role as a carbon sink, and the carbon pumps will start to sputter. We will not be able to resolve everything because many effects have been irreversibly set in motion for the next few centuries. These include increased water temperatures, rising sea levels, weakening circulation, acidification, and oxygen loss. But rather than a reason for despair, it is a reason to reduce our CO_2 emissions as quickly as possible, for every tenth of a degree counts. In reality, emissions should go all the way down to zero. We should also drastically reduce our emissions of other greenhouse gases, such as methane.

Another essential measure is to protect the ocean and all its biodiversity. Only 2.7% of the world's waters are actually *really* protected (on paper, 7%). When protected areas recover, they will be better able to store carbon. The same applies to coastal biomes such as mangroves, giant kelp forests, salt marshes, mud flats, and seagrass meadows. These ecosystems lost half of their acreage in the previous century to human pressure, rising sea levels, warming, and extreme climate conditions. Turning this evolution around and taking advantage of these ecosystems' strength is not only good for our climate but also for coastal protection, biodiversity, and the fishing industry.

The humble North Sea can perhaps also contribute to the battle against climate change. Key measures include offshore renewable energy, storing carbon below the seabed, and storing 'blue carbon' in ecosystems and marine reserves. Yet, we should not be foolhardy and rely too much on the latest technological developments. They are promising, but a great deal of scientific research is still needed to determine their feasibility as well as their ecological benefits and drawbacks. Think about the heated debates around the effects of wind farms on ecosystems, for example.

Geo-engineering entails even more risks. There are several options. Some suggest fertilising the sea with nitrogen, phosphorous, and iron to encourage the growth of phytoplankton to increase CO_2 absorption. Preliminary tests have not proven very successful thus far. Other suggestions that sometimes pop up in the press: injecting liquid CO_2 into deeper water layers; cultivating and sinking algae mats for storing carbon; dissolving olivine, which binds easily to CO_2, in coastal areas; and reflecting sunlight through ice, foam, clouds, or other means. This all sounds very promising, but history has taught us that meddling with complex systems can lead to unexpected consequences. And not all of them are always positive. With many of these methods, furthermore, we do not know whether they will be sufficiently effective in practice and at a relevant scale. The research is often fantastic and very laudable, but we should not be lulled into a false sense of security by depending on solutions that will still take many years to implement. We do not have much time, and we already have enough insight and technology to start taking effective action now.

The two terms that are currently very popular in that respect are 'mitigation' and 'adaptation'. Mitigation is important because every tenth of a degree that we can avoid counts. If we are realistic, we have to realise that climate change cannot be stopped. Adaptation, therefore, is the logical next step: we need to be prepared and adapt. Think of measures such as protecting our coasts and water supplies. The ocean can play a crucial role in climate change, hopefully for the better. However, for the biological pump and other mechanisms to do their work, the ocean needs to be healthy, which means we will need to tackle other problems as well. Pollution is one of the obstacles that we need to overcome, and it is certainly not the easiest.

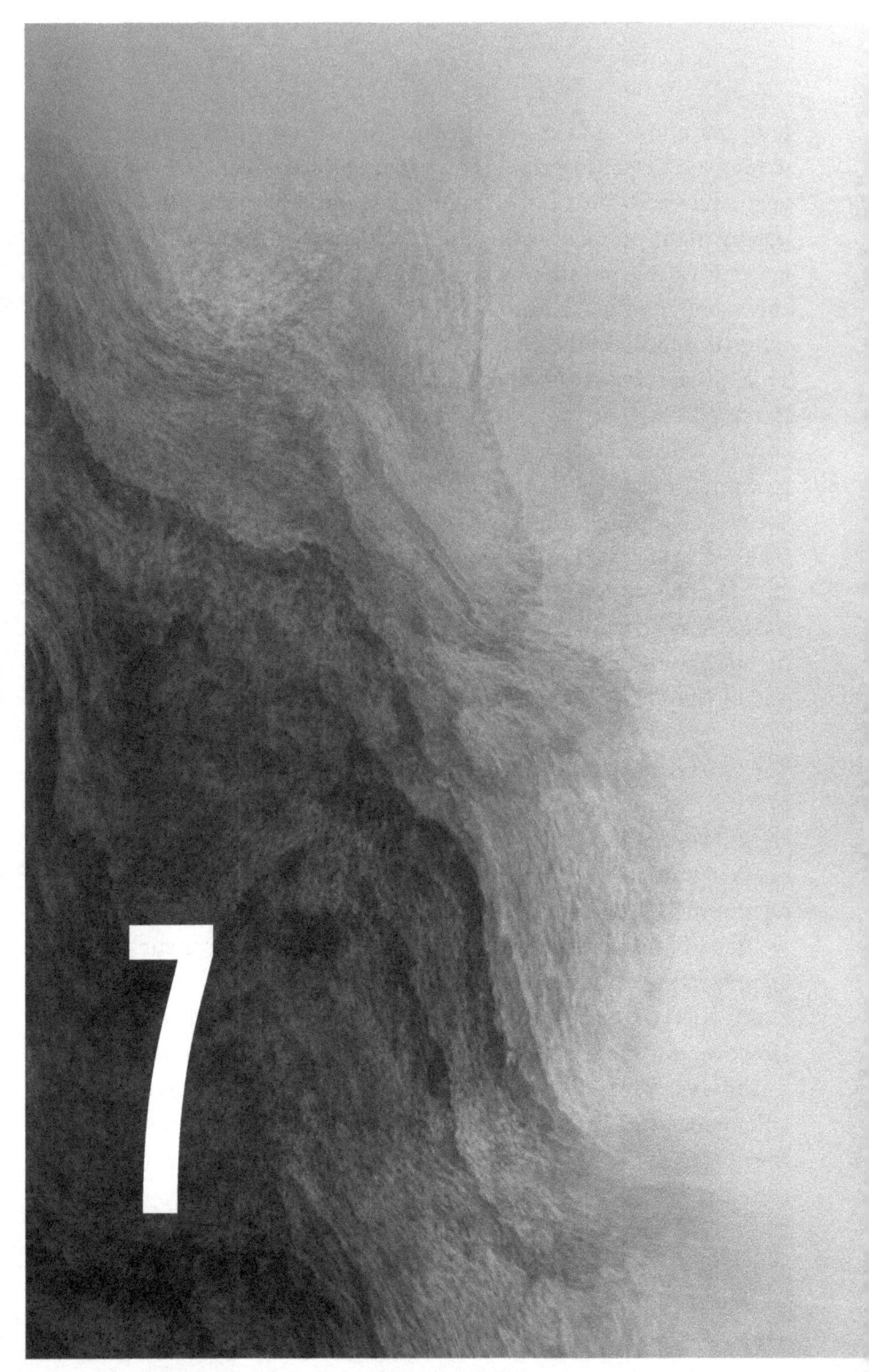

POLLUTION, A MULTI-HEADED MONSTER

> "The sea told me that she is tired today
> That she is not feeling well, in disarray
> It said: help me understand
> What are you all doing on land?
> It said: there are days, more and more,
> That I can't stand your filth and gore."
> **PAUL VAN VLIET, DUTCH COMEDIAN AND SINGER**

In the late 1960s, as soon as I (Colin) would come home after a day at the beach, the same daily ritual would take place. My mother would come up to me with petrol – she would later use white spirit – to wash the black globs of goo from my trousers, boots, and hands. How could I help it that my number-one playground was riddled with oil and tar?

During those same childhood adventures, I discovered another odd thing on the beach: I saw foam, lots of foam. In my memory, it actually towered above my small figure. For a very long time, I thought it was detergent that had foamed up because of the crashing waves. I saw similar foam floating in canals and creeks, and people told me that laundries and other industries caused it. Later, I learned that I was wrong. What causes sea foam is the brown alga *Phaeocystis globosa*, which we discussed earlier in our exploration of the North Sea. The proliferation or 'bloom' of this alga was the result of another form of pollution: eutrophication (an excess of nutrients flowing into the sea from the land). The crashing waves whip up the proteins and sugar chains in the foamy algae, which sometimes makes the sea look like a giant cappuccino.

In those long-ago summers, we would often swim near the fisherman's quay, where the tourists and all kinds of other folk washed up on shore. That was never a good idea. The area was often signposted with 'No swimming' signs, and for good reason. Wastewater from the sewer system, teeming with germs, flowed from the city straight into the sea through a drainage pipe that, to a ten-year-old, looked suspiciously like a secret tunnel.

These are my memories and, at the same time, three examples of what would later become the subject of my research: the pollution of the sea.

THE SEA IS TIRED

Entire libraries have been written about the many aspects of pollution and its impact on habitats. It is impossible to provide a comprehensive overview. Therefore, we will limit ourselves to a brief introduction to this highly complex topic, focusing on human involvement and the historical evolution of marine pollution. On the one hand, we see significant improvements;

on the other hand, recent problems such as plastics and microplastics pose new challenges that are not as easy to address. We want to show how our way of life and the way in which we interact with the sea directly or indirectly negatively impact our health.

Just like climate change, pollution threatens the stability of essential processes on our planet, leads to loss of diversity and freshwater reserves, and threatens our food supply and many other benefits provided by ocean ecosystems. Pollution today is a major environmental cause of disease and accounts for millions of premature deaths each year. It also causes tremendous economic losses. It may sound strange, but until recently, marine pollution was overlooked by the international organisations that deal with the global economy and healthcare. For a long time, pollution was considered an inevitable side effect of economic growth, and people thought that combatting it would stifle progress. We now know better: many countries that have doubled their gross national product in recent years were also able to drastically reduce pollution. This means that progress is no longer an excuse for inaction.

For many years, the general approach to marine pollution followed the old saying: 'The solution to pollution is dilution.' In other words, the ocean is so big that everything is watered down so much that it no longer causes any harm. Many assumed that humans were incapable of harming the sea. Again, we know better now. Initially, ocean pollution looks a lot like land pollution: a complex, ever-changing mix of human-produced chemicals and biological materials, including plastic waste, petroleum-based pollutants, metals, industrial and household chemicals, pharmaceutical products, pesticides, and a toxic cocktail of nitrogen, phosphorous, artificial fertilisers, and wastewater. We can best visualise this pollution cocktail as an iceberg that only has its tip protruding from the water, with the bulk of the pollution literally hidden below the surface of the sea, either in solid form or dissolved.

Some substances have been in the ocean for a long time, while others are new and largely unknown. Some substances also break down more quickly than others. Marine bacteria, for instance, can break down many but not all of the substances found in household wastewater. In some cases this process is very quick and in others very slow. The same applies to river-borne wastewater that reaches the sea from agricultural lands and the food industry, breweries, paper factories, and other sources. Bacteria can even combat oil pollution but, as we have mentioned above, this statement needs to be qualified with a huge 'in theory' because bacteria need time. Moreover, bacteria can only process so much of these substances.

Paradoxically, other types of pollution are the result of too many nutrients: eutrophication. Nitrogen and phosphorous can feed phytoplankton in the sea to the extent that some species become dominant and overrun everything. The result is red, green, and brown algae blooms that can cause oxygen deficiencies and release hazardous toxins. These toxic substances have detrimental effects on the organisms living in coastal ecosystems, such as the fish population, but they can also be harmful to people living on the coast, causing health problems via contaminated air and seafood (for example, diarrhoea or paralysis from ingesting mussels that feed on toxic algae). The unchecked growth of large algae through eutrophication and the washing up of vast swathes of seaweed on our shores add additional problems to the mix: stench, beach-going or boating experiences that are less than pleasant, and clean-up costs.

Far more worrisome are all those substances that are hard or impossible to break down. They can do quite a lot of harm. Examples include heavy metals, DDT and other pesticides, antibiotics, and PCBs. There are simply too many to mention, and every year a couple of hundreds more are added to this toxic cocktail. Sometimes we do not even know which compounds are toxic and what their risks are for humans and the environment. When you realise that some 350,000 different chemicals are used worldwide, many of which have barely been researched, you cannot help but worry. Chances are that the discussions we are currently having about PFAS will rapidly shift to other substances as soon as science has caught up to the industry.

Other forms of pollution have only recently been discovered, such as human-made underwater noise. Noise can significantly impact marine life, including whales, dolphins, seals, and other species that use sound to communicate, navigate, and forage. Sources of underwater noise pollution include shipping, drilling and construction at sea (such as driving piles for wind farms into the sea floor), military sonars, and underwater explosions. The effects of underwater noise can be both physical and behavioural, and include impaired hearing, relocation, changes in migratory patterns, and decreased reproduction levels.

Unfortunately, as we will see later, new forms of pollution are still being added to the mix, but thankfully, much has also changed in the past 20 to 30 years. We are now more environmentally aware of what happens on land and in our waterways. The problem is that the focus of this concern and awareness has not yet been placed on the ocean, where it is just as sorely needed. People sometimes still have the idea that the sea is big enough to sufficiently dilute our waste to render it harmless or that we can allow it to accumulate without risk. However, many scientific studies conducted over the past 50 years prove that this is an incorrect assumption. To come back to the quote of Dutch comedian and singer Paul van Vliet: the sea is tired.

FROM OIL POLLUTION TO PLASTIC SOUP

The older generation will perhaps remember the *Amoco Cadiz* oil spill. This oil spill took place on 16 March 1978 off the coast of Brittany, France. The oil tanker ran aground and split in two, spilling about 220,000 tonnes of crude oil into the sea. The result was one of the most catastrophic oil spills in history, with devastating effects on marine life and the environment. The oil soon washed up on shore and polluted some 320 kilometres of coastline, affecting tens of thousands of birds, sea mammals, and fish. The outcry from the general public was huge, and the spill led to stronger regulations on oil transport to prevent future spills from occurring.

Here is another example from our collective consciousness: 20 April 2010, when one of the biggest natural disasters in recent history occurred in the Gulf of Mexico, not far off the coast of Louisiana. Deepwater Horizon, an oil rig leased by BP for drilling oil, exploded. Eleven people died instantly from the explosion, and countless other effects were felt in the weeks, months, and even years that followed. It is the largest marine oil spill in history. The striking images speak for themselves:

birds so covered in oil that they can barely move and cleaning them is almost impossible. Dead fish wash up on shore, and vulnerable ecosystems such as coral reefs and river estuaries are thrown off balance and need to recover from the disaster. More than a decade later, scientists are still studying the spill's consequences for marine life in the Gulf of Mexico and documenting its adverse effects on ecosystems.

Such disasters are spectacular examples of ocean pollution. Thankfully, they are quite rare. Although the local damage they cause is enormous, the global effects of these incidents are fortunately limited. What most people do not know is that most of the oil in our ocean does not come from oil spills and other shipping disasters, but from other sources. About 5% comes from natural sources, and 35% comes from oil tankers and other shipping activities, including illegal dumping and tank-cleaning operations. Oil spills account for only a small fraction of the total volume (5–10%, depending on the period and the source). Oil also enters the ocean through the atmosphere: volatile oil particles that are released into the atmosphere during combustion end up in the water. These particles in the atmosphere, together with pollution through municipal and industrial wastewater and oil from oil rigs, account for about 45% of oil pollution. Another 5% comes from 'undefined sources'.

And yet, there is less oil pollution than there was, say, about 50 years ago. Where 79 oil spills occurred in the 1970s, only 6 took place in the 2010s. Children seldom come home from a day at the beach nowadays with trousers soiled with thick black globs of crusty oil. Illegal dumping is more difficult these days because our waters are more closely monitored than previously. Governments guard the coastal waters with ships, aeroplanes, and drones. When dumping violations occur, we can determine which ship the oil comes from based on water and oil samples.

We are now more concerned about the combustion of oil and natural gas, which is responsible for air pollution and the emission of greenhouse gases. As for ocean pollution, the oil problem has made way for another petroleum-derived product: plastic. That plastic is everywhere now, and it is at least as visible as oil. Future archaeologists will say that we lived in the age of plastic. This form of pollution leaves few people indifferent when they see bottles, plastic packaging, or other plastic waste floating in the sea or lying on the beach.

PLASTIC'S TRIP AROUND THE WORLD

It could have been a beautiful and inspiring story, with a gifted Belgian-born New Yorker in the leading role. In 1907, Leo Baekeland developed the first fully synthetic plastic, the forerunner of all plastics today. The substance was given the name Bakelite. It was an incredible invention that triggered the start of the plastics industry, an evolution that would fundamentally change the world from the 1950s and 1960s onwards. This is a product that is relatively easy to produce and has countless applications: it is an ideal construction material and an ideal packaging material for food and pretty much anything that has to be transported – plastic rules everywhere: on the land, in the sea, and even in the air.

Unfortunately, over time, it became apparent that its usefulness was also its biggest drawback. We find plastic literally everywhere. Slowly, technological optimism gave way to concerns about a global problem.

In 1997, the American oceanographer and yacht captain Charles Moore made a striking discovery. Upon his return from a yachting race in the Pacific Ocean, he spotted a massive field of plastic waste floating somewhere between Hawaii and California. News headlines soon talked about the discovery of a plastic island in the middle of the Pacific Ocean that measured about a million square kilometres. It is still referred to as the Great Pacific Garbage Patch. Just to be clear: it is not a real island; you cannot actually walk on it. It is a constantly shifting zone that changes size and position depending on the ocean currents, which explains the great variation in estimates of its size. The 'island' does not have any clearly defined borders, but compared with other zones, it clearly has much more waste.

Recent studies show that the ocean contains about 171,000 billion floating plastic particles (or larger pieces). Depending on the source and the estimate, they amount to between 80,000 and 269,000 tonnes of plastic waste. If we take the highest estimate, we can compare this weight to that of roughly 45,000 African elephants (about 11% of all African elephants on Earth).

The Great Pacific Garbage Patch contains about 29% of all floating plastic in our oceans. And according to recent studies, that amount is increasing exponentially. Where does this plastic come from? About 75% of the larger plastic pieces come from lost and dumped fishing gear (cords, nets, and plastic tubs) from fishing vessels originating in Japan, South Korea, China, the US, and Taiwan.

Why is the waste concentrated in this region? To address that question, we need to refer back to the ocean currents. Wind and surface currents guide the plastic towards giant oceanic whirlpools called gyres, where it tends to accumulate. One of those gyres happens to be in the exact location of the Great Pacific Garbage Patch. But there are two more in the Atlantic Ocean, one in the Southern Pacific Ocean and one in the Indian Ocean, and you will also find massive accumulations of floating plastic there.

Every year, millions of tonnes of discarded plastic find their way into the ocean. Some estimate that the equivalent of a full bin lorry is dumped into the sea every minute. It is hard to fact-check such an estimate because bin-lorry sizes are not standardised. The estimates also have a margin for error, so depending on how you set upper and lower limits, every year between 4.8 and 12.2 million tonnes of plastic end up in the sea. Regardless of which estimate you adopt, it is an enormous amount. Part of it floats around in giant rubbish dumps or washes up on the shore, but most of it sinks and lands on the sea floor. Another part breaks down and turns up in food such as mussels and fish. Whatever we throw into the sea eventually ends up on our plates, as we will see later.

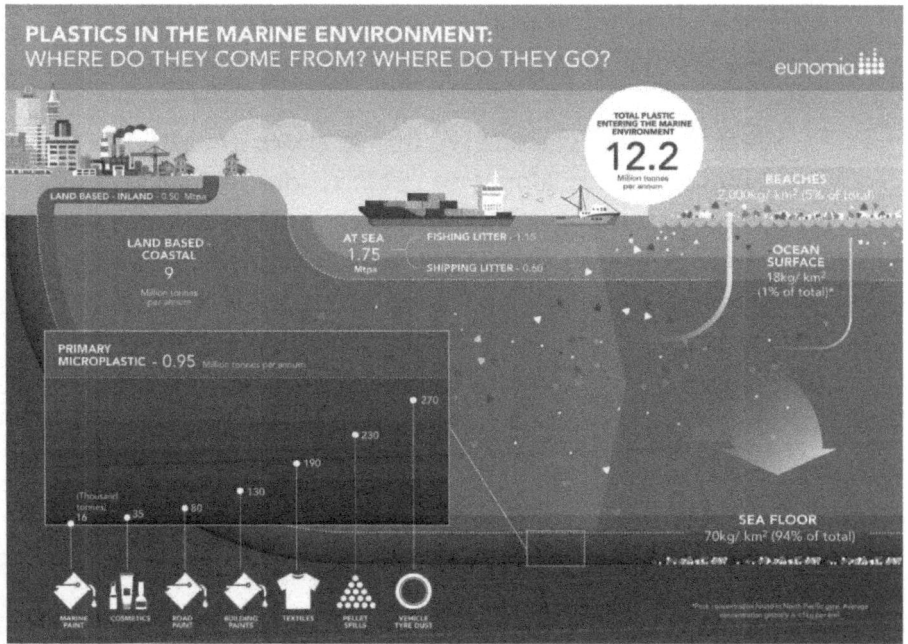

Figure 9. How much plastic ends up in the sea every year, where does it come from, and where does it go? Source: Eunomia Impact Report (2016)

To understand where all that waste comes from, we need to look at the source of the chain. According to the Organisation for Economic Co-operation and Development (OECD), we produce about 460 million tonnes of plastics each year. These figures date from 2019, so the number may be larger today. Four hundred and sixty million tonnes is the equivalent weight of about half a million Eiffel Towers, although this comparison hardly helps paint a concrete picture because of the sheer unimaginable volume. Of course, these estimates have their margins of error, but still, we produce *a lot* of plastic. Moreover, the last ten to fifteen years have seen exponential growth, with the world's production in 2010 reaching 210 million tonnes per year.

What is all that plastic used for? In short: everything, in every industry, and in almost every household. Packaging material accounts for 42%. There is much ground to be gained here in the fight against plastic waste. The construction sector takes second place with 19%. The plastic used in the packaging industry is also less sustainable: it has an average lifespan of 6 months, while the plastic used in construction averages 35 years. But plastics are also used in clothing (think of textiles containing polyester, nylon, or acrylic), cosmetics, toys, plastic bottles, toothbrushes, smartphones, television sets... The number of practical applications is mind-boggling.

Of all the plastic produced since 1950, more than half has ended up in a rubbish tip or has been thrown away somewhere. Only 6% has been recycled, only 20% of which was still in use in 2015; the rest has been burned or dumped. Recycling is better than nothing, but it is grossly overrated: most of the plastic eventually ends up being some form of pollution, even if it is initially recycled. That is why the saying today goes: *refuse, reduce, reuse & recycle* – in that order. First, we need to avoid plastics. Only when this is not an option should we look to reduce, then see whether they can be reused or recycled.

Let's turn back to the sea. How does all that waste eventually end up in the water? More than 80% of all plastic in the sea comes from land, just like 80% of all ocean pollution in general originates on land. Twenty per cent comes from 'marine' sources such as fishing nets, fishing lines, and abandoned boats. A recent study shows that more than 75% of the plastic in the plastic soup of the Pacific Ocean Garbage Patch comes from fishing. But generally, the problem mainly originates on land, from rubbish, waste falling out of overfull rubbish bins, single-use bottles, and plastic bags. Most of this plastic makes its way via the wind, sewer systems, and canals to rivers, which eventually flow into the sea.

Although the situation is dire elsewhere in the world, the rivers that carry the largest volumes of plastic to the sea are found in Asia. Seven of the ten most polluted rivers are found in the Philippines. Together, they are responsible for no less than 12% of global plastic pollution in the ocean. The combination of large coastal cities, poor waste management, and fast-flowing rivers is particularly disastrous. In comparison, Europe does quite well: 0.6% of all plastic in the ocean comes from European rivers. Although we have to be fair and say that Europe sometimes exports its plastic waste to other countries, where it still ends up in the sea, so the figures would probably be much higher in practice. There is no reason to feel proud.

How bad is all that plastic? The effects on marine life are nothing to be sneezed at. We are all familiar with pictures we have seen in the media: dead whales on the beach, their stomachs filled with dozens of plastic bags and even plastic jerry cans; seabirds that have eaten plastic or are trapped in fishing wires or cords; sea turtles dying from plastic in their intestines or unable to reproduce; and so on. These animals often mistake plastic for food. Sea turtles eat plastic bags because they mistake them for jellyfish. Some birds, such as the albatross, are drawn by the smell of microalgae living on the plastic. Recent studies show that almost 700 species of marine animals suffer from plastic pollution, and that number seems to be growing.

Moreover, new challenges arise when plastic starts to break down without actually disappearing: the notorious microplastics.

MICROPLASTICS

Captain Charles Moore's famous sea voyage caused a lot of commotion in the late 1990s. Suddenly, the media was filled with reports about plastic pollution, and slowly the magnitude of the problem dawned on the general public. It helped that the Great Pacific Garbage Patch was highly visible, tangible, and 'mediagenic'. Still, Moore was not the first to raise the issue. Research had already been conducted into plastics and their effects on marine life in the 1970s and 80s. Even back then, scientists found phytoplankton containing microplastics: tiny plastic particles (defined as 5 millimetres or less, but often much smaller). They did not, however, realise the full implications of their discoveries at the time. Only in the early 2000s did research into microplastics really get started.

Today, we know that the microplastics found on beaches come from many sources: all kinds of paints, care products such as cosmetic scrubs and toothpaste, as well as textiles and the plastic resin pellets or tiny spheres that are used in the production of industrial plastics. But the wear and tear on tyres – our cars and lorries – also contributes to the supply of tiny plastic particles in the sea.

Perhaps microplastics also provide an answer to 'the mystery of the missing plastic'. There is a considerable discrepancy between the amount of plastic that winds up in the sea and the recent estimates of plastic floating in the ocean. The balance does not add up: we do not know where more than 90%

(some even say 99%) of the plastic that enters the sea eventually ends up. We could simply conclude that our estimates are wrong, but a popular theory holds that ultraviolet radiation and waves break plastic down into ever smaller pieces: microplastics. Those particles find their way into sediments or are ingested by animals, including one of Belgium's coastal delicacies: mussels. Everywhere we look, we find microplastics: on beaches and in the deepest deep sea; in sea salt, sea ice, and sea air; in plankton and bottom dwellers; and in fish and sea mammals.

Our own story about plastic waste and microplastics began in 2005 at Ghent University. Seeing the newspapers full of headlines about the plastic soup and plastic islands, we wondered about our own back garden, the North Sea. How much plastic, big and small, lies on our beaches, off our coast, and on the seabed? Back then, very few people in Europe were researching that topic, so there was very little reliable data. Just like many other countries, the Belgian government promised in all sorts of international forums that it would protect the sea, but we found that there was no scientific literature about microplastics in Belgium. That is why we can consider the beginning of microplastic research at Ghent University – almost 20 years ago now – as pioneering work that acted as a catalyst for worldwide research into microplastics.

How do you research microplastics? To start with, you need to do what countless children do at the beach every summer: dig holes and shovel sand. But instead of building sandcastles, we collect samples. Then we isolate the plastic. You can try this yourself, provided you live or are somewhere near the sea. You take a bucket of wet sand and add plenty of salt. You stir the sand or shake the bucket thoroughly, and then you leave it to settle. Most of the plastic will float to the surface of its own accord. After that, it is just a matter of filtering the microplastics out of the surface water and placing them under a microscope.

When you add dye, you can recognise the plastic particles, although there is quite a significant margin of error using that technique. Later, we used a hot needle to test for plastics: if the material burns, it is probably organic, and if it melts, then it might be plastic. Meanwhile, we use advanced equipment to count the number, type, and size of the microplastics in a sample. We were among the first to look for microscopic particles, pushing the boundaries of light microscopy: you cannot get any smaller than that unless you use an electron microscope. The next step was to start collecting and analysing data from North Sea beaches. From an international perspective, this was groundbreaking research. One of our first conclusions: microplastics are everywhere, although they are more common near ports and harbours. We even found 100 times more invisible plastic than visible plastic. For every kilogram you see, there are 100 kilograms you do not see. Plastic breaks down into smaller pieces. It does not decompose in the sense that it disappears; it just gets smaller.

In fact, we do not know exactly what the lifespan of plastic is. You may have heard or read that a plastic bag can last up to 20 years and that a straw may take 200 years to break down. However, those figures are not based on solid empirical research. Ultraviolet radiation can break down plastic's polymer chains, but that does not apply to the deeper, darker regions of the ocean.

We also took sand samples that showed the evolution of pollution over time, similar to ice cores or soil cores that show the evolution of climate change. We found that the bottom sand layer contained only a third of the amount of plastics in the top layer. This gave a clear picture of the accumulation over time: 30 centimetres is a little less than 50 years. You can clearly see how the problem has grown since the 1970s.

The research does not stop there. We also want to know the effects, both for marine life and for us humans. The problem was that most studies were carried out in laboratories that exposed test organisms (such as mussels and small crustaceans) to unrealistically high plastic concentrations. The results from laboratory experiments using concentrations that are 10,000 to several million times greater than the concentrations we actually find in the sea say very little about the actual environmental effects. We felt that we could do better and adopted an entirely different research approach: we let the marine organisms tell their own story. We looked for microplastics, particles measuring between a couple of dozen to hundreds of micrometres, in the bodies of organisms measuring a couple of centimetres. After many experiments, we eventually developed a technique for dissolving all the body tissues of mussels and other marine animals in acids and other chemical compounds. What was left after that process were the plastics.

We studied both mussels and lugworms; one filters water and the other eats sand. They represented two different ways (water and sediment) in which marine organisms are exposed to plastics. The spread was relatively even: microplastics were found in all the mussels caught along our shores. Later, we conducted the same experiment in other European countries and came to the same conclusion. Plastic is everywhere. The same results were found for the worms. To find out where the microplastics were located in mussels, we 'purged' the mussels for our analysis: we cleaned their intestines. This enabled us to establish that the microplastics were located in the actual muscle tissue, thus preventing the possible criticism that plastic can be removed from mussels by 'diluting' them (placing them in clean water to get them to release sand). Based on these findings, we calculated that a regular mussel eater ingests up to 11,000 microplastic particles each year through consuming seafood.

Not surprisingly, the above-mentioned criticism, which we were able to refute immediately, came from mussel farmers in countries known for their large and tasty mussels, such as Belgium and the Netherlands. Because of this, our research garnered quite a bit of media attention. We were suddenly *persona non grata* to the mussel farmers. Some of them defended their product by saying that they 'diluted' their mussels, a treatment that causes them to release their waste products. Home cooks may have heard of the traditional wisdom of cleaning mussels in salt water so they release their sand and grit. That works well but does not remove the microplastics from the mussels' tissues. After receiving this criticism, we went to a local fishmonger to buy mussels from Zeeland and applied the same technique, obtaining the same results. We also conducted CT scans, which showed that the plastics had accumulated in the mussels' hepatopancreas, which means they had worked their way through the animals' intestinal walls. Two different techniques with the same result.

Our research even became news in Switzerland, a country that is not particularly known for its mussels because it is landlocked. They also carried out studies on shop-bought mussels, with the same results. It is quite a sensitive issue: when the Swiss came to Zeeland to see what was what, they almost got chased away by the mussel farmers, who are not really the culprits but the victims of pollution, just like the mussels themselves.

So now we know that marine animals store microplastics in their tissue. Those plastics can damage organs, reduce reproductive capacity, transfer toxins, and have a detrimental impact on other animals further up the food chain.

What does that mean for human health? That is a question that has not yet been fully answered, but we have found some pieces of the puzzle. We were among the first researchers to study the effects of eating fish and seafood. We know that microplastics can penetrate human intestinal cells because we cultivated them ourselves. Meanwhile, we also understand that the problem goes beyond fish and seafood. Microplastics enter our bodies just as much via other foods and drinks as through the air we breathe. We are constantly breathing in microfibres, for instance from clothing or synthetic carpets. Laboratory experiments have shown that the substances in plastics can influence our nervous system and hormones.

To be completely certain, we need to do more research. Meanwhile, how do we deal with this uncertainty? Are we allowed to eat mussels and oysters, or better not? How much is hazardous? Those are difficult questions and unfortunately we do not have a simple answer, but we may find part of the solution in the principles of risk management and in advice from a 16th-century Swiss doctor.

PARACELSUS' WISDOM

There is much confusion about what the word 'pollution' means in everyday language. Technically, it means the introduction into the environment of a foreign substance or contaminant, potentially causing exposure to something harmful. Both elements of this definition require some explanation. The words 'foreign substance' or 'contaminant' raise the question of to what extent something is considered 'unnatural' or 'foreign'. That is a difficult question to answer because what *is* natural, strange, or foreign? In the case of plastics, it is pretty clear. But what about oil, which is, technically speaking, a natural product?

The potential exposure to something harmful involves two related concepts. In English, we can distinguish between 'hazard' and 'risk'.

The term 'hazard' involves an element of danger. It is something dangerous and likely to cause damage. That could be a chemical substance, a machine, a virus, or a natural disaster. A hazard can have a direct or an indirect detrimental impact on humans or the environment.

The second term we need to define in this context is 'risk'. Risk is the likelihood that a particular danger (hazard, in other words) in a certain environment will actually lead to damage or loss. So, there is a quantifiable probability factor involved that lies somewhere between zero (no chance) and one (absolute certainty). Not only is the source of the potential damage critical in this respect, but also its potential impact – the consequences, in other words. Let's illustrate this idea with an example. Corrosive acids are intrinsically dangerous (hazardous) for the skin. But as long as they are kept in a secure container, the risk of harm is low.

How can we mitigate the risks to our health and the environment? One way is to adopt environmental standards. Governments and international organisations set these standards in collaboration with scientists and other experts. The determination of such environmental standards is usually the result of a scientific process involving several stages.

The first step is identifying the environmental problem that needs to be addressed. Then, data is collected about exposure to and the effects of the levels of the substances and activities that may cause these environmental problems. That data is then analysed to assess the risks to human and environmental health. After that, the government or international organisation defines the standard and translates it into legislation. Practical and political feasibility plays a key role in this process. Finally, inspectors monitor adherence to the standards. The enormous distances at sea make everything more difficult than on land, especially in areas outside of national jurisdictions.

It is not always easy to determine what is hazardous in practice. A 16th-century Swiss doctor saw the reason for this problem coming. He was born with the impressive name of Philippus Aureolus Theophrastus Bombastus von Hohenheim, although he personally probably preferred his self-chosen nickname: Paracelsus. His most famous quote is, 'All things are poison, and nothing is without poison; the dosage alone determines that a thing is not poison.' Or put more succinctly: 'the dose makes the poison'. Even water can be deadly if we drink it in large quantities for extended periods. A substance's toxicity, therefore, depends not only on the nature of the substance but also on the concentration and the duration of exposure.

(Eco)toxicologists attempt to consider these factors when determining environmental standards. They try to find a relationship between the concentration of a substance and the health of the animal, plant, or ecosystem in question. This is usually done in a laboratory environment, by exposing model organisms (usually animals) to increasing concentrations of a substance to ascertain its effects. Based on these observations, researchers can establish dose-response relationships, which refer to the relationship between the concentration and the severity of the resulting health effects, as well as the threshold at which a response is observed.

Test subjects do not always have to be typical lab animals such as mice; different types of tissues and cells are also suitable for testing. For research into ocean conditions, a variety of organisms such as fish, crustaceans, molluscs, and algae are often used. For example, scientists can test the effect of algae-repellent coatings on algae populations. Ideally, we would then know the safe concentration of the tested substance.

Applying this to plastics, we can conclude that we know the *hazard* part of the pollution equation from laboratory research. We know that plastic can be hazardous both to us and marine life. But to what degree and at what level of exposure? That is the *risk* part, and that factor is still largely unknown. However, we also have something called the precautionary principle: if we know that something is potentially harmful, we must limit exposure as much as possible. Or as much as is desirable, in any case. The precautionary principle is actually a form of risk management and, therefore, finds itself on the exceptionally complex interface between science on the one hand, and politics and applied ethics on the other. It is a cost–benefit analysis with a massive dose of uncertainty. How careful you *must* be is often also the subject of academic debate. There is no scientific method that can answer that question because the 'must' part implies an ethical consideration rather than a scientific one. We scientists can, at best, give as accurate advice as we possibly can while remaining transparent about what we do not know. In doing so, we should not only be aware of the dangers posed by the fact that we have incomplete knowledge but also of the fact that dangers may crop up that we are not aware of in the first place. Donald Rumsfeld, the former US Secretary of Defence, coined the notorious terms 'known unknowns' and 'unknown unknowns'.

Sometimes, we know that we do not yet know something, like the exact effects of plastics on health. However, there could also be things that are not acknowledged as potential dangers simply because we have not thought of them yet. A striking historical example is the Minamata disease or 'dancing cat disease' that we discussed in Chapter 5. Of course, politicians also need to consider socio-economic factors. It is one thing to say that we should eat fewer mussels, but that can hardly be considered welcome advice for the coastal economy. Moreover, many people from vulnerable regions around the world are highly dependent on the sea for their food; they simply do not have the option of basing certain choices on health considerations.

What is clear, no matter what angle we look at this from, is that we need to do more to combat plastic pollution – and, by extension, all pollution. And that is our task as scientists: to tell people where we stand.

A SILVER LINING ON THE HORIZON?

Fortunately, there is hope. Recent studies have shown that we are more careful with plastics. Before 1980, burning and recycling plastic was practically non-existent: almost 100% went to rubbish tips or was thrown away. In the 1980s and 1990s, we saw that more and more plastic was being burned or recycled. If we extrapolate that evolution, we should arrive at the following results: by 2050, the burning of plastic will rise to 50%, recycling will reach 44%, and only 6% of plastic waste will be thrown away. Still, we should be cautious when making simple forecasts like this.

Moreover, we should not rely solely on technological solutions. Sometimes, sensational articles appear in the media about enzymes that can break down plastics or about how Boyan Slat has found a new method of extracting plastic waste from the Great Pacific Garbage Patch with The Ocean Cleanup project. But remember that, even if they *do* work, projects like these impact only a tiny fraction of the total amount of plastic in the ocean, while huge quantities are being added every day. It is literally just a drop in the ocean. It also gives the wrong impression, making it seem like there are easy ways to solve these problems. The same applies to bioplastics. Remember what we said about the degree to which something can be considered 'natural': whether you make plastic from oils or straight from plants, it does not make much of a difference. Biodegradability tests are carried out under unrealistic conditions, at 60°C with enough oxygen. These circumstances rarely occur naturally.

The solution is that we need to produce and use less. We are living beyond the capacity of the Earth to sustain us. We need to dare to be more radical. What we do see is that bans on single-use plastics work. At some point, policymakers also made the radical decision to say things like: no more lead in petrol. This decision met much criticism at the time, and many said that it would never work. But eventually, it became the new normal. The same could also apply to plastics and other forms of pollution. Let's hope that we act quickly and finally put the commitments we have made on paper into practice.

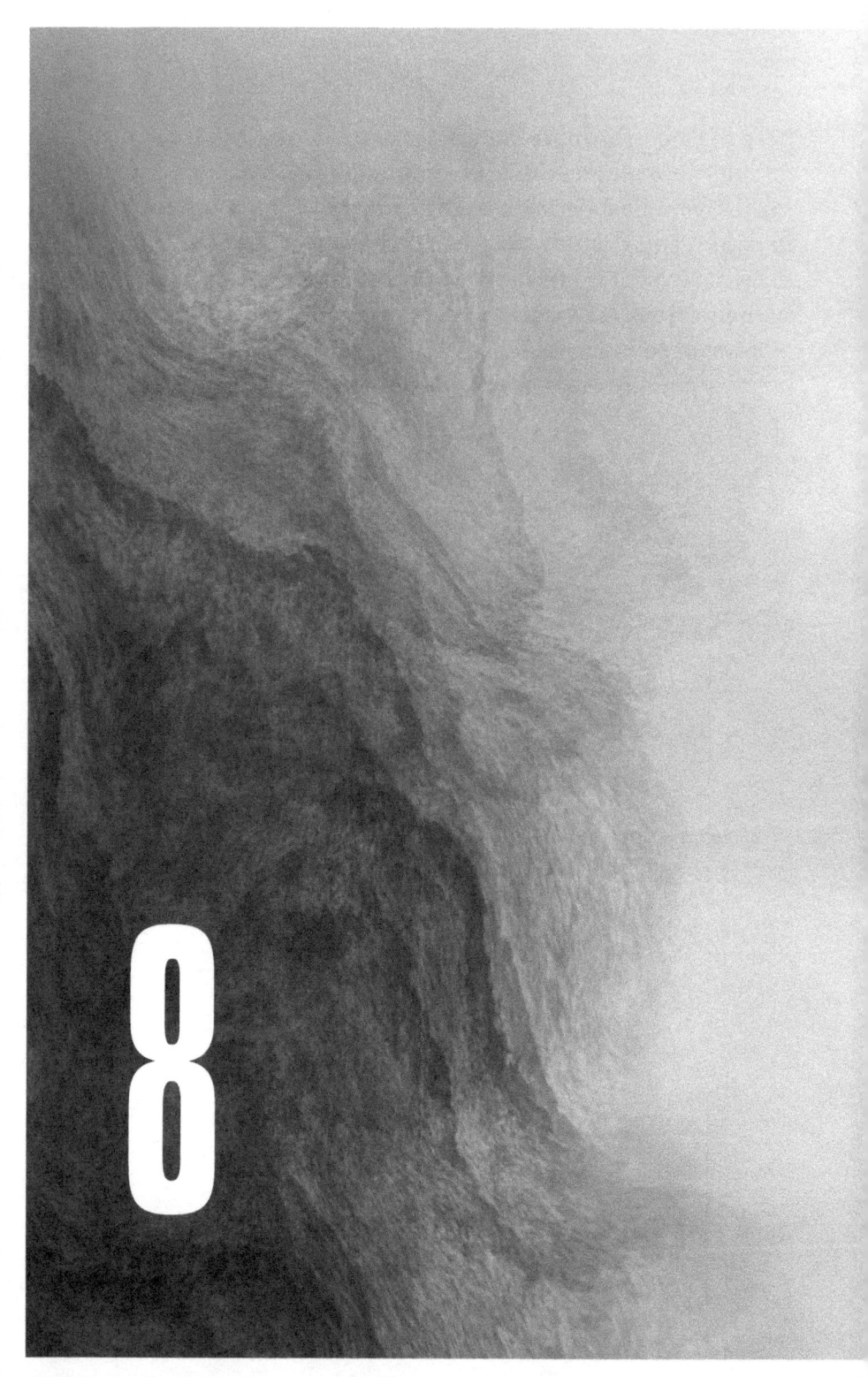

ARE THERE ENOUGH FISH IN THE SEA?

> *"Down in the ocean, where all the fish roam free, you'll look for and find a meal that happens to look quite a lot like me."*
>
> **MESKEREM MEES ('BLUE AND WHITE')**

Anyone who has spent prolonged periods on the turbulent high seas can probably imagine the romanticism of life as a fisherman. In popular television series or documentaries, fishermen are often portrayed as sea dogs with a heart of gold. Rugged men – there are very few female fisherfolk – hit the decks through wind and weather, clad in yellow oilskins, while others on board gut the fish with sweat running down their faces. Meanwhile, they must weather storms and the occasional illness: fishermen on the high seas cannot just go home. Teamwork is essential to bringing the voyage to a successful conclusion. It is hard work, but the freedom, companionship, and rewards give the profession an allure that many modern-day office jobs lack. Intriguing sea shanties, often accompanied by pounding rhythm, add to the romanticism. In early 2021, when COVID-19 still had the world in its grip, a sea shanty went viral, a new recording of 'The Wellerman', an old song about whaling.

But times have changed, and over the past few decades, small-scale local fishermen have had to compete with large fishing fleets that overfish the seas with their dragnets and use sonar to chase down the last fish. These fishing fleets are not necessarily made up of large factory-type boats; even the 'normal' coastal fishing boats can severely damage fish populations and their habitats, especially if they engage in bottom trawling. The romanticism of traditional fishing has lost much of its charm as fish populations decline and sustainability decreases. Meanwhile, the industrial fishing fleets destroy what they want to achieve: enough fish to eat. In short, the system has derailed.

This tale does not have to end badly because we can still turn the tide. And the solution for restoring marine biodiversity is deceptively simple: stop destructive industrial fishing practices. Whale populations are living proof that this works: they recovered as soon as whaling was banned. The blue whale once faced extinction at the hands of humans, but its numbers are increasing. After many decades of exploitation, the whale hunting recounted in the sea shanties of old is as good as gone following a worldwide moratorium and protective measures. There are also other success stories where fish populations have recovered after their spawning grounds or nurseries were designated as protected areas.

SEASPIRACY

The issues associated with overfishing became known to the general public with the release of the popular documentary *Seaspiracy*, which premiered in 2021. In this documentary, the young photographer Ali Tabrizi researches the impact of fishing on the ocean and humans. It was a great success: the documentary gained a spot on the top-10 list of the most-watched programmes on Netflix. The producer is Kip Anderson, who also released the notorious and hotly debated documentary *Cowspiracy* about the negative impact of intensive agriculture.

Seaspiracy's message is simple: the ocean is on the verge of collapse, and it is high time we did something about it. The main culprit is the large-scale fishing industry, which often sets its own destructive course in unmonitored and immensely remote parts of the ocean. The essence of the message is correct: the oceans are suffering from industrial-scale overfishing. And the solution is indeed simple: we must kill fewer fish. While nothing new for scientists, until then this was not well known to the general public. In that respect, the documentary has been valuable.

Tabrizi, however, paints a very ugly picture of the fishing industry, and experts were quick to criticise the documentary for its lack of nuance and many inaccuracies. Eminent marine scientists such as Daniel Pauly and Ray Hilborn, who generally disagree with each other on overfishing, were in total agreement that the documentary did more harm than good. Their criticism was that anyone who uses poor argumentation, tendentious claims, and erroneous facts for a good cause is not helping that cause. Bryce Stewart, a marine ecologist, voiced a similar concern on Twitter (now known as X). 'This is the worst kind of journalism,' he wrote. 'People will either believe it completely and overreact, or find it so easy to discredit some of the statements that the real issues get downgraded/disbelieved.'

The documentary's claim that the oceans will be empty of fish by 2048 dates back to a scientific paper from 2006. The statement in question was included in the accompanying press release and then took on a life of its own. The actual conclusion was far more nuanced: by 2048, fish stocks will only generate 10% of what the maximum yield was in 1960 – which is a serious problem in itself, of course. Fish populations continue to plummet worldwide, and yet we continue to fish unabatedly. If we do not cut back on our fishing, this could have serious consequences, initially for the fishing industry itself and later for all of us. But we will not leave our oceans empty.

This might sound like the argument that fishermen often use: the sea is too big for overfishing. But just because our seas' fish stocks cannot be completely depleted does not mean that rampant overfishing will not harm the ocean's health, our food supply, marine biodiversity and, through the disruption of ecosystems and the biological pump, even the climate.

The documentary also fails to show the benefits and successes of fisheries management. Sustainable fishing is a viable option when we implement restrictive measures such as fishing quotas. These types of measures have been proven to work for various species in North America and Europe.

What the documentary does point out, and quite rightly so, is that governments support overfishing by going against scientific advice and allowing more fishing than is prudent. That happens year after year in Europe as well; a situation which will ultimately negatively impact the fishermen themselves. Moreover, governments subsidise the fishing sector, leading to a considerable overcapacity in the fishing fleet: they can simply catch far too many fish and easily overfish the already limited fish stocks. It is an industry that is kept afloat by unnecessary amounts of money. There are even subsidies for what we call IUU (illegal, unreported, unregulated) fishing and for the use of fossil fuels. In 2023, the EU accepted the World Trade Organization agreement on fisheries, which states that subsidies to non-sustainable fishing practices must be stopped.

Tabrizi also targets sustainability labels such as Dolphin Safe and the Marine Stewardship Council (MSC), claiming that they are no good. We can and should be critical of them, but they are still often the lesser evil if we, as consumers, want to continue to eat seafood and want to have a say in how fish are caught and farmed. Tabrizi has the wrong enemy in his sights: the fishing industry is doing the overfishing, not the NGOs that assign the labels. Those labels are a form of consumer pressure and help people to make better, more conscious choices. In that respect, they are helpful in the necessary battle against overfishing and non-sustainable fishing practices.

We consumers can, to a certain degree, rely on sustainability labels, but that does not mean that they make the problems go away. One example is the lobster fishery in the Gulf of Maine, certified by the Marine Stewardship Council (MSC). This fishery only exists because the area's ecosystem had completely collapsed due to the earlier overfishing of cod. The lobsters, which were eaten by the cod before the ecosystem collapsed, were able to thrive in a strongly downgraded and vulnerable ecosystem. So, the MSC label gives us the wrong message here.

The final destination of fisheries' products should also be considered. For example, MSC certifies certain fisheries where the catch is ground down into fish meal, which is used as fertiliser or as feed for pigs and chickens on land, or in aquaculture cages for salmon. That jars. This usually involves factory ships that catch small pelagic fish such as anchovy, mackerel, horse mackerel, herring, sprat, and sand eel. Between 1950 and 2010, about 20 million tonnes of fish were caught each year worldwide for non-human consumption.

Sometimes, we even fish lower down the food chain, which we will come back to later. The recent certification of Norwegian fishing vessels, licensing them to catch krill in pristine Antarctica, is inexcusable. The krill is reduced to fish meal and serves as food for intensive salmon farming.

On the other hand, small-scale, artisanal fishing deserves better protection because it feeds people directly (short supply chain), creates more jobs, and improves sustainability (less by-catch).

Tabrizi's main conclusion is 'stop eating fish'. In doing so, he conveniently ignores the fact that millions of coastal residents in less-developed countries are dependent on their daily catch to get enough protein or scrape together their daily wage. Or as Daniel Pauly, who we will meet again several times in this chapter, said in an interview with the news website Vox: '*Seaspiracy* undermines its tremendous potential value: to persuade people to work together, and push for change in policy and rules that will rein in an industry which often breaks the law with impunity.' According to Pauly, the way forward is a sustainable and strongly regulated fishing industry. It is primarily governments that make the decisions which affect the ocean's well-being; 90% of the worldwide fish catch is determined by just 30 countries and the European Union. Stopping fishing altogether is not realistic; we need to focus on what is feasible. As the old Italian proverb goes, 'perfect is the enemy of good'.

SCALING UP

Fishing has been around for as long as humans have existed. Even the Neanderthals used to fish, as we discovered from the remains of fish bones and shellfish in a cave in Portugal. These protein-rich fruits of the sea were a part of their diet. Overfishing has also been around for a very long time, albeit on a more local scale. We know this from Aristotle's writings, in which he points out overfishing in the famous Bay of Kalloni in Lesbos, Greece. Fishers scraped the seabed in search of scallops, causing the molluscs to disappear from the bay. The Romans may also have been responsible – in addition to 'regular' overfishing – for the extinction of several whale species in the Mediterranean, depriving the sea of nutrients and further decreasing the fish biomass (the total amount of fish) in the region. Whales are incredibly important for ecosystem productivity, in part through the nutrients in their faeces.

However, vast stretches of the sea were once inaccessible to humans because we simply did not have the ships and the technology to fish there safely. Until the Industrial Revolution, fishing was primarily restricted to coastal waters. Storms and wars sometimes hindered further exploitation of the ocean, but it was mainly its inaccessibility and dangerous depths that held most people back.

The Industrial Revolution accelerated technical progress and the scaling up of industry. From the 1870s onwards, fishermen could drastically increase their catch with engine-driven vessels. They also had to become more productive to afford all this expensive new technology. In addition, the fishing industry was already starting to decline in both Western Europe and the United States, so fishermen had to scale up their investments to maintain their catches.

Industrial fishing started in 1877 with the first steel steam trawlers cleaving the waves in the United Kingdom. Fishermen could fish further, deeper, and longer. Within 20 years, coastal fish populations had been decimated, and the capital- and energy-intensive fleet moved its activities out to the open sea. It was the start of the globalisation of industrial fishing, driven by a pattern of collapse through overfishing and the consequent depletion of fishing grounds through geographical expansion. Later, we shall see why this pattern has similarities with a pyramid or Ponzi scheme.

Catch sizes increased until they rapidly declined from the 1950s onwards. You would think that catches would increase with the growth of the world fleet's capacity and improved technology with stronger ships and sonar. But the fact that it did not was bad news. Above all, it meant that the effects of overfishing were becoming apparent.

At the same time, fishing fleets were scaling up: smaller boats were having an increasingly tough time. In Belgium, as in most European countries, the entire sector has been in decline since the 1970s: the number of ships, the number of fishermen, and the catch size decreases every year and continues to do so today. Fish populations are slowly being exhausted.

There are a few success stories for some species that are closely monitored. But for most fish, this is sadly not the case. Fishing is currently the greatest force exerting pressure on marine ecosystems worldwide, more than climate change or pollution. And just to be clear, we are not talking about small-scale local fishing, but about large-scale industrial and bottom-trawling fishing practices, not to mention illegal fishing techniques such as dynamite fishing. The upshot is that fish populations are collapsing and habitats are being destroyed, creating ecosystem-wide effects that we often do not yet fully understand but which may be irreversible.

A well-documented example is the Canadian cod-fishing industry along the Atlantic east coast. Despite a complete ban since 1992, fish populations have still not recovered. This suggests that the ecosystem has arrived at an alternative stable state dominated by different, smaller species that are less interesting to the fisheries. We do not yet know if recovery is possible in all the other documented cases of overfishing because the fishing continues unabatedly. Examples are global problems with sharks, rays, certain tuna species, and the Peruvian anchoveta, as well as trawl fishing in the Gulf of Thailand, recent problems along the west coast of the United States, and closer to home, the general malaise in the Mediterranean and the depressing state of cod populations in the North Sea. However, we should generally remain optimistic: in most cases, populations will probably recover when given the chance.

HOW MANY FISH LIVE IN THE OCEAN?

You sometimes hear, 'There are plenty more fish in the sea,' in response to a broken heart. Sadly, this does not apply to actual fish. But how do you count how many fish there are?

The global Food and Agriculture Organization (FAO) of the United Nations keeps statistics on the worldwide fish catch. It has been doing this for decades. Each year, countries report on how much and which species of fish they catch, the techniques they use, and the regions where the fish is caught. Data is also collected on aquaculture (the farming of fish, algae, molluscs, and shellfish). This sounds easier than it is because many countries have trouble collecting accurate data. There are many uncertainties involved, but we are very aware of these uncertainties – they are known unknowns – and the general trends are clear.

We produce a total of about 214 million tonnes of fish and seafood each year. China is the largest producer by far at 60 million tonnes, followed by Indonesia, India, Vietnam, and the United States. A striking point to note is that global fish production has increased fourfold in the past 50 years. The world population has doubled, and the average person eats twice as many fish as before. Aquaculture barely existed 50 years ago, but today it has become more important than wild capture.

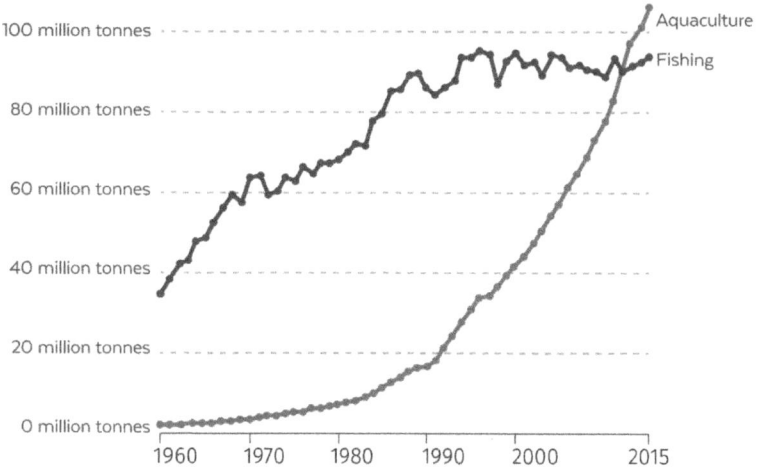

Figure 10. Wild capture has stabilised since 1990 around 90-95 million tonnes per year, while production from aquaculture has increased to 100 million tonnes per year over the same period. Source: from FAO (2022) and https://ourworldindata.org/fish-and-overfishing.

As far as wild capture is concerned, the statistics confirm the overfishing narrative. Wild capture has stabilised since 1990 at around 90-95 million tonnes per year, while production from aquaculture has skyrocketed from less than 20 million tonnes to more than 100 million tonnes per year over the same period. This is disconcerting: we have not been able to catch more fish in over 30 years while worldwide fishing capacity has increased rapidly in that time. That is a clear sign of overfishing. Moreover, if we look at the state of our current fish stocks (a stock is the 'harvestable' part of a population, and a population is, in turn, a biologically distinctive part of a species), we notice something else. We see that the number of overexploited fish stocks has more than doubled since 1980. In other words, we can catch fish faster than they can grow and reproduce to maintain their numbers.

But are these figures accurate? As we said, there is a huge margin of error on the figures supplied by individual countries. There are also a couple of interesting phenomena behind the figures, as we shall see.

Let's first address the discrepancies in the statistics. Daniel Pauly published information about these statistical errors as early as the turn of the century. Most countries are simply unable to collect and pass on reliable data. That is because there is a lot of illegal, or at least unreported and unregulated fishing out there. Not all fish that are caught end up in the statistics; in reality, the numbers are much higher.

Additionally, many fish are thrown overboard as bycatch and die in the process, others disappear on the black market, and others still are 'uncountable'. By 'uncountable', we mean sectors such as recreational fishing or small-scale local fishing in countries with extensive coastlines and many landing points. Some countries also have trouble documenting how many fish are taken from their waters by foreign fleets. As a result, most countries under-report their catch. Pauly and his team have attempted to reconstruct the statistics for each country with estimates for the missing catch.

Another error in the statistics that may sound counter-intuitive is that some countries tend to over-report their catch. That has been the case in countries where you can further your career by meeting predetermined targets and saying how well your fishing sector is doing. Sometimes, figures from certain countries had to be removed from analyses because they were completely unreliable.

After careful adjustment of the figures from the Food and Agriculture Organization, it turns out that the situation is worse than the official statistics had us believe. The peak in the global catch a couple of decades ago was actually much higher (130 million tonnes around 1996), and the stagnation that we saw at the end of the 20th century turned out to be a decline that continues to this day. The effect of overfishing is, therefore, greater than we initially thought. But that does not stop us from trying to catch more, which means that every year, we tap into new areas and kill too many fish. Those 130 million tonnes are estimated to encompass 20% of the fish biomass found in the ocean at any given time. For sustainable fishing management, you can catch roughly 10% of the biomass. Everything above that figure is overfishing. This figure also does not include important side effects such as the excess bycatch of other species, the damage to and destruction of habitats (such as mussel banks and cold-water reefs), genetic changes in species, climate effects caused by disturbing the seabed, and many others.

TECHNOLOGICAL CREEP, FOOD WEBS, AND SHIFTING BASELINES

The figures do not look good, even if we consider all the uncertainties. However, there are also several underlying trends we need to consider: technological creep, fishing lower down in the food web, and shifting baselines.

We have already seen an example of technological creep in our story about scaling up. We are slowly getting better at catching fish. It used to be that we were limited by natural factors such as geography and the seasons during which ships could not sail because of stormy weather, or because fishermen knew they would lose their nets to stones and shipwrecks in certain regions. Technological progress has allowed us to overcome those barriers. Ships have more powerful engines, travel further, spend longer periods at sea, and drag heavier, stronger nets. We now have modern navigation techniques and sonar, improved weather forecasting, supporting charts, and mathematical models. The fish do not stand a chance: we can hunt every last fish or school down.

Fishing boats travel more often to other areas. Fishing fleets, including western fleets, sail further to areas that were once rarely fished, if at all. Sometimes, they travel to tropical regions such as the exclusive economic zones of African countries that do not have the means to fish, manage, or protect their 'property'. Sometimes, the ships travel to the continental slopes and submarine mountains of the deep sea and the recently discovered twilight zone. Declining catch numbers in traditional fishing grounds are compensated by fishing new stocks and species elsewhere, which are consequently overfished. This trend is, in essence, the result of fleet overcapitalisation and the perverse subsidy system: those ships have to keep fishing to turn a profit, so they travel deeper and further.

There are countless recent examples of such expansions. We have already seen how krill is being fished in Antarctica to make products that include food for our pets. In 2020, a large Chinese fleet of some 300 fishing boats was discovered fishing off the Galapagos Islands, known for their exceptional biodiversity since Darwin's journey of discovery on the *Beagle*. The public outcry was just as great as the number of squid and other marine animals the boats hauled aboard. In 2022, when a US Coast Guard ship went to inspect if there was any illegal fishing going on in the region, it put geopolitical relations between China and the US even further to the test. We can quote many more examples of such expansion. Industrial fisheries also scour the coastal areas of Latin America and Africa.

The composition of the catch is changing as well. Because the larger fish are depleted first, we often see a shift to smaller species. Within the catch, there is a shift from apex predators at the top of the food chain, such as tuna and cod, to smaller prey. This phenomenon is called 'fishing down the food web', another term coined by Daniel Pauly and his colleagues.

We know that populations of larger species such as sharks and rays have decreased by 70 to 90% of their 'virgin' biomass. Tuna and related species had dropped to between 40 and 50% of their original population in 2010. A recent study on the decline of the iconic sawfish, also known as carpenter sharks, shows that they have become extinct in the waters of 55 of the 90 countries where they once occurred. Cod, an apex predator, has almost disappeared in our waters. So, we fish increasingly smaller fish, which keeps catch sizes steady but masks critical effects on ecosystems. These smaller fish are often not suitable for human consumption but are processed into fishmeal that is used as cattle feed, fish feed in fish farms, or fertiliser. Constant improvements in fleet technology do not lead to greater catches of apex predators – you cannot catch what is not there. In fact, often fewer fish are caught, especially when we are talking about structural overfishing, as was the case with cod. The overcapacity is then used to catch economically 'inferior' fish. Note that fishermen do not always fish lower down in the food web because the higher levels are depleted, nor is higher up the food chain synonymous with greater value. The lower ranks of the food chain can also be highly profitable; take shrimp as an example. However, the average trophic level – the rank in the food chain – of catch worldwide and in a variety of regions seems to be decreasing over time.

Part of our reluctance to do anything about overfishing stems from industry lobbying, although that is not the only reason. There is an interesting phenomenon that alters our frame of reference, the 'shifting baselines'. This is yet another term from Daniel Pauly. You have probably guessed by now the seminal influence of this French Canadian scientist on research on overfishing.

What is shifting in those baselines? The idea of what a 'normal' catch should be, for one. Younger generations of fishermen have an increasingly low estimate of what constitutes a 'good' catch. That may sound like good news on the face of it, but the reason behind it is more nefarious: overfishing. In a recent Belgian television series, *Een jaar op zee* (A Year at Sea) by Wim Lybaert, an old fisherman that fished in Icelandic waters in the 1970s tells us of boats filled to the brim with giant cod. They even had to work extra hard to find enough room to stock all the fish before taking it back home to Belgium, often discarding 'older' fish for freshly caught, more valuable ones. You may still occasionally hear such stories, but the reality has changed fundamentally. Today, a catch of ten cod would be considered a successful haul by a coastal fisherman. And those fish are much smaller than they used to be. You might be tempted to think that those stories about incredible catches are exaggerated, but historical photos and film footage clearly illustrate how richly stocked the North Sea and the Mediterranean were in the first half of the 20th century. Large sharks and rays swam everywhere, and fish markets were well supplied with shiny tuna and giant cod.

The Mediterranean Sea is known today as a relatively poor sea. That is probably because we have no recollection of what swam in its waters a hundred years ago, let alone in ancient times. A good catch for the Greeks would probably have been different from what the Romans thought and anyone who came after them.

So, overfishing is nothing new, just like deforestation, which has been around for centuries – another not so positive legacy left by the Romans in Europe. For too long, we believed that the ocean was too big for us as humans to have any significant impact. Even in the early 2000s, representatives from the Belgian Shipowners Association claimed that overfishing did not exist. The sea was inexhaustible. The Belgian fishing industry, however, was floundering at the time: there were fewer than 130 ships, and people were saying that the system was on the verge of collapse. That was the context in which those claims were made.

But we knew better, even then: Daniel Pauly published his controversial studies showing that the global fish catch was declining in spite of improved techniques and more fishing. Now, twenty years later, we see few signs of improvement. The Belgian fleet has shrunk to 60 ships in 2024, a large part of which is in foreign hands. At a European level, we still see the same mechanism through which the ministers of fisheries, year after year, ignore the scientific recommendations from the International Council for the Exploration of the Sea (ICES), raise the catch quota, and facilitate overfishing practices for another year.

SOLUTIONS FOR OVERFISHING

Humanity constantly has to draw on its ingenuity to think up solutions to problems of its own making. For the overfishing issue too, several interesting avenues of exploration have been suggested.

In theory, *Seaspiracy's* recommendation of not eating fish is an option, but it is not feasible in practice, especially on a global scale. About 15% of the proteins that people ingest come from the ocean. In some countries, this percentage is much higher: about a fifth in Japan, Iceland, and Cambodia, and up to a third in island states like the Maldives. There are also huge differences in consumption: in Eastern European countries, the average person eats about 6 kilograms of fish per year, while in China and Iceland, it is 40 and 90 kilograms respectively. Many European countries, the US, Canada, Russia, and Australia are situated somewhere in between at 20 to 25 kilograms per year.

It is also a social problem. Wealthy countries like ours could perhaps stop fishing altogether, but that is not an option for many regions in the world where economies and people are dependent on seafood. That should be kept in mind too. In any case, small-scale fishing in coastal areas needs to be protected from industrial fishing.

It would also help to close the high seas to fishing, as Daniel Pauly also suggests. The catch there only accounts for about 5 to 6% of the total catch and is mostly for the rich Western countries, so imposing a fishing ban in those regions should not be a problem.

Fish should not be underestimated as a source of food. And remember, by 2050 we will have 10 billion mouths to feed (our planet is now home to 8 billion people). Researchers predict that global food production will need to increase by over 50% to meet the nutritional needs of humankind. Intensive agriculture is already under fire for its adverse effects on climate, land use, water availability, and biodiversity. And when we look to the sea, we find that most fish stocks have been fully exploited or overexploited.

That is why aquaculture, mostly of fish but also the farming of invertebrates such as crustaceans and shellfish, aquatic plants, and algae, offers an interesting alternative. We should, however, be aware that this also has limits. There is already insufficient space for current aquaculture practices in coastal areas, leading to conflicts with other users. A strong expansion could potentially lead to other major environmental issues: deforestation (especially harming the 'blue carbon' ecosystems), eutrophication, and the spread of diseases to populations in the wild.

Sixty per cent of the global production from aquaculture comes from one single country: China. And it is mostly freshwater. The ecological impact is mixed. The production of algae and shellfish could be sustainable, provided that not too much coastal habitat is destroyed to make room for it. Coastal areas protect us from the effects of climate change, and mangrove forests, seagrass meadows, and salt marshes are often breeding grounds for fish that are essential to local fishing. One aquacultural practice that will definitely not help solve the problem of overfishing or the world's growing food needs is growing carnivores such as salmon because they need fishmeal to grow. For every kilogram of salmon, you need between

2.5 and 4 kilograms of other fish captured in the wild. It is much more efficient to eat those fish directly.

So, if we want to use aquaculture as a solution, we need to shift from predatory fish at the top of the food chain, such as salmon, to animals that are further down. An obvious candidate would be algae, which have already been proven to play a crucial role in the sea on multiple fronts, as we showed earlier.

Algae are highly diverse; we have already mentioned that algae is an umbrella term for different types of organisms: bacteria, archaea, and eukaryotes. They are also rich in nutrients, containing vitamins, minerals, antioxidants, omega-3 fatty acids, and more proteins and amino acids than land-based plants. Their cultivation can also help combat pollution, in combination with waste management from fish farming. Microalgae-based products also have great potential for the production of additives and as substitutes for meat, dairy, and baked goods. Their productivity is much higher than that of land plants too.

Naturally, we should not pin all our hopes on aquaculture and algae. We first need to ensure that current fishing practices become more sustainable. And there is already movement in that direction. A relative success story in terms of sustainability is the Alaska pollock, massive numbers of which are used for surimi (imitation crab meat), fish burgers, and other products.

The underlying principle is that we can keep fish populations up indefinitely as long as we give them enough time to recover each time. If they are sufficiently allowed to grow and reproduce, fish can theoretically serve as a renewable food source. You can probably feel a 'but' coming: we still have a long way to go in practice. The fishing industry remains focused on maximising profit. When you combine this with efficient but disruptive fishing techniques such as dredging or bottom trawling – dragging nets over the ocean floor – many ecosystems become severely damaged.

In Europe, fisheries management is still a matter of assigning quotas, with too little attention to overcapacity, destructive fishing techniques, perverse subsidies, developing truly protected areas (net-free fishing zones), ecosystem recovery, and gaining fundamental knowledge about the animals we exploit.

We believe that drastic measures are needed. Since these measures will have to be taken at some point, we should do so sooner rather than later.

To start with, we urgently need marine protected areas, reserves where fishing is not allowed. On 4 March 2023, the United Nations signed a historic treaty: 30% of the ocean in areas beyond national jurisdiction must be protected by 2030. But what happens in practice remains to be seen. In many of our current marine reserves, even industrial bottom trawling is still allowed, so we cannot really say that these areas are protected. It would help if there were more reserves, because even the smallest reserves make a big difference. They also benefit neighbouring areas. We are not saying that these areas should be permanently closed off. Clearly demarcated regions such as reserves are highly effective, but so are temporary 'no-take' zones, where fishing is prohibited in certain zones at certain times of the year when the populations are vulnerable during reproduction and growth.

The industrial fishing sector must change. It is simply not sustainable, not even in economic terms. According to Daniel

Pauly, the model behind it is similar to a giant Ponzi scheme: as soon as one area is emptied of fish, you move on to the next. At the same time, new fishermen join in while the fishing regions themselves thin out. It is a finite system if you do not give the fish stocks time to replenish. That is why we need to stop subsidising industrial fishing practices. Every year, the sector receives 20 to 30 billion dollars in funding, mainly to help cover fuel costs. The result is overcapacity. If this funding disappears, boats will no longer fish where profits are low. It will probably also solve the waste problem (every year, 10 million tonnes of fish are thrown back into the sea as bycatch) and the fact that a third of the catch goes to animal and fish feed. Small-scale local fisheries are much more sustainable than giant trawlers that deplete life on the sea floor. The consequences of this last practice are clearly visible in China, where small-scale local fishing has been practised for thousands of years. Today, fish stocks for them have dropped to 16% of what they were before the Industrial Revolution.

In fact, all destructive fishing techniques should be banned everywhere. Dredging or bottom trawling, fishing the sea floor with drag nets, is a clear example. The consequences not only affect fish populations but also biodiversity in general and even our climate. Everything is linked together, as we have indicated throughout this book. Thankfully, initiatives have already been implemented on that front: in the EU, bottom trawling has been prohibited at depths greater than 800 metres since 2016. What is absurd is that it *is* still allowed in 'protected' areas. So, we should take this term with a huge grain of sea salt. Discussions are underway about prohibiting destructive techniques in protected areas too, but it is a slow and difficult process.

You can also help as a consumer by eating less fish that is under threat (such as eel and bluefin tuna) or caught unsustainably (such as certain stocks of cod and sole). It is not always easy to determine, but online consumption guides posted by organisations such as the WWF and the Marine Conservation Society (MSC) are there to help. They do not agree on everything, and we have already discussed some valid criticisms of these labels, but they do help us get a general sense of what is what. For instance, you can find out what fish species are caught using disruptive bottom-trawling techniques or come from overfished regions.

THE WHALE AS A TALE OF HOPE

In an ideal world, marine animals themselves would be better protected, and it would be prohibited to feed fish and invertebrates to other animals. According to Daniel Pauly, we should view marine animals more as wild animals than as commodities.

There are examples of what that would look like in practice. One example is whale fishing, which we briefly referred to at the beginning of this chapter. The International Whaling Commission was established shortly after the Second World War. The initial goal was to protect commercial whaling because this sector was under threat from declining populations due to overexploitation. In the beginning, whales were treated as fish and managed as such by assigning quotas, closing off areas and implementing seasonal closures, prohibiting fishing of certain species or of calves and their mothers, and so on. These measures were not enough; whale populations continued to fall. In 1982, when thanks to the efforts of the great

British biologist and founder of fisheries science Sidney Holt, whales were beginning to be seen as intelligent beings with their own culture, a moratorium on whaling was issued. Commercial whaling was completely phased out by 1986. During the meeting on that moratorium, Holt said: 'It is a great evil to destroy something we do not understand.' There were exceptions to the moratorium. Since 1986, Japan has been allowed to capture a couple of hundred whales each year 'for scientific purposes'. An exception was also made for traditional hunts by indigenous peoples around the Arctic Circle in Alaska, Norway, Greenland, Russia, and a few Caribbean islands with a whaling tradition.

The moratorium was a success, after which some members saw no further reason to limit themselves. In 1993, Norway and Iceland resumed commercial whaling with self-imposed quotas. In 2014, the International Court of Justice ruled that Japan was abusing its right to exemption and that it had to stop whaling altogether. But to this day, no majority has been reached for ending the moratorium, although there is also no consensus on turning the moratorium into a permanent ban. In the Florianópolis Declaration of 2018, the commission's purpose was adjusted to the conservation of whales and the safeguarding of marine mammals in perpetuity. All whale populations must recover to pre-industrial whaling levels. Japan withdrew its membership later that year, announcing that it would resume commercial hunting within its territorial waters and exclusive economic zones (the zones in which a country has exclusive rights regarding economic activities up to a 200 nautical mile limit off the coast) but that it would cease whaling activities in the southern hemisphere.

The protection of whales did not happen through gradual adjustment and improvement in management practices but through improved scientific knowledge, increased global awareness, the adoption of the prevention principle, and a generous dose of political courage. The result: new laws, new social norms, and a new relationship with the animals themselves. Whale populations are slowly recovering.

What we were able to do for 90 whale species, we should also be able to do for the thousands of other marine animals that we consume as food. This is how we can make fishing sustainable. Then, we can once again proudly talk about our fishing culture with its rich traditions and romantic songs without having to refer to the elephant in the room – or in keeping with our current theme, the whale in the swimming pool: overfishing.

On a final note, our fight against overfishing should not be seen in isolation from climate change and pollution. When water warms, it contains less oxygen, which is exactly what fish need to deal with the rise in temperature. The result? Fish are getting increasingly smaller. This is actually one of the consequences of the pressures of fishing, which acts as a force of 'natural' selection, pushing fish species to reach reproductive maturity at an increasingly earlier age and smaller size. The biological pump will also not work as efficiently as biodiversity changes drastically, and pollution makes life harder for a variety of organisms. This shows us that the ocean's three major threats – climate change, pollution, and overfishing – are linked and should be addressed as a whole. Adopting this holistic, integral approach will require the help of new scientific and technological developments, but we will also need to change our behaviour.

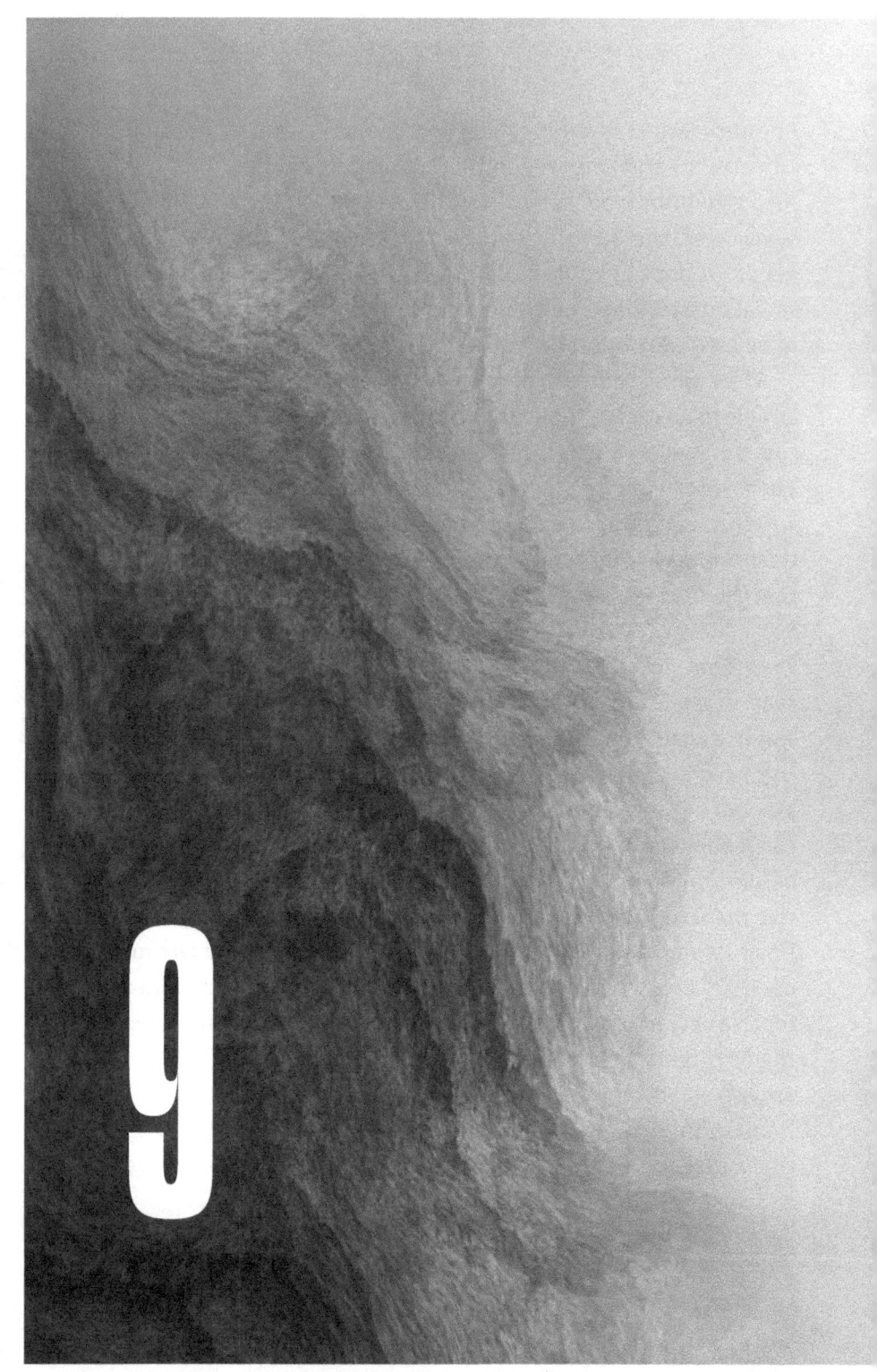

THE BLUE ACCELERATION

> "*The Ocean Is the Ultimate Solution.*"
> **FRANK ZAPPA, SLEEP DIRT**

While the deep, dark sea remains mostly a mystery, it is slowly starting to reveal its secrets. We have made countless fascinating discoveries, of which we have discussed just a few, such as giant squid, exotic microbes, and life in inhospitable places where no one imagined life was even possible. These discoveries also include new resources in our waters, such as manganese nodules: potato-shaped, metal-rich rocks that lie on the seabed. For quite some time, we have also known that the ocean harbours many riches, such as oil and gas. The temptation to exploit these resources is great, especially now that the demand for raw materials used in electronics and other products is growing. Moreover, the technologies used to harvest them from the sea are getting better.

Does the future of humankind lie in the ocean? This chapter aims to provide a nuanced answer to this question. It is an urgent question because, as we speak, economic pressure on the sea is mounting due to oil and gas extraction, shipping, and industrial fishing – all activities that exploit finite resources and maximise profits to the extent that they end up destroying those resources, eventually leaving everyone worse off.

Still, other developments are taking us in the right direction. There are more and more wind farms, which may have detrimental effects on some birds and other animal species, but which generate renewable energy and make us less dependent on fossil fuels. Theoretically, they can mitigate the effects of climate change and reduce the threats to marine life.

THE HOCKEY STICK

Such complex and often contradictory developments we summarise under the term 'blue acceleration' because this acceleration is the common denominator of all these developments. It reminds us of the Olympic motto *citus, altius, fortius* (faster, higher, stronger), where the Latin *altius* can refer to both 'higher' and 'deeper', which is more appropriate when talking about the ocean environment. If we translate all these different developments into a scientific graph, we end up with one that, with a bit of imagination and the use of a seemingly very non-scientific term, looks like a hockey stick. It is a type of figure that shows fast, sometimes exponential growth at the end, similar to the graphs you see on climate change. Figure 11 compares a variety of developments: aquaculture, oil and gas exploitation in the deep ocean, deep sea mining, the desalinisation of seawater, the use of marine genetic material, shipping, submarine pipelines and cables, cruise tourism, wind turbines at sea, the expansion of coastal areas, but also the number of oceanic observations and the ocean volume marked as protected marine reserves.

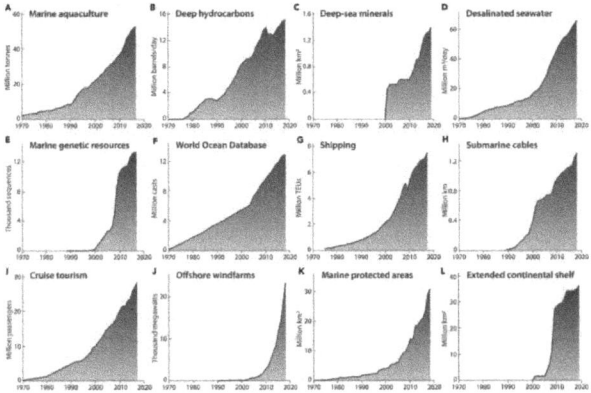

Figure 11. The blue acceleration: these are just a few examples of accelerated growth in marine resources. Source: Jouffrey et al. (2020)

Blue acceleration also includes positive developments. Examples include an increased focus on the ocean at prominent events such as climate and biodiversity summits, the declaration of the Decade of Ocean Science for Sustainable Development by the United Nations, and the recent UN treaty around the protection of areas beyond national jurisdiction. Even the ocean's economic exploitation is not as black and white as it seems. For instance, we need metals to produce chips for electric cars, which in turn makes us less dependent on fossil fuels and their harmful emissions.

All this is happening at breakneck speed. The problem with such rapid growth is that we could face potentially enormous effects, and we do not know whether the course we are setting out on is the right one. If we compare our knowledge of the ocean to a ship at sea, then it is a vessel that needs holes to be plugged while sailing on the open ocean. What are the large-scale effects of ocean acidification, overfishing, and loss of biodiversity, for instance? What are the dangers of micro- and nanoplastics, pesticides, and countless synthetic substances? We know enough to warn people of the dangers, but we only know a tiny fraction of the potential consequences. Meanwhile, industrial expansion and development continue at lightning speed.

IS THERE ENOUGH ROOM IN THE OCEAN?

One of the consequences of the blue acceleration is that we are taking up an increasing amount of space in the ocean. Think of aquaculture farms, offshore platforms for oil and gas, wind farms, and similar coastal and offshore structures.

Another key player that is taking up ever more space is transportation. For millennia, the sea served as a medium for disseminating ideas, power, and trade, and it became a catalyst for globalisation. However, the introduction of container shipping in the late 1960s revolutionised oceanic transport. Today, more than 80% of world trade takes place by sea. Think about all of those packages that you order online: many of them are shipped by sea.

Even the seabed is slowly becoming human territory. An impressive 1.3 million kilometres of submarine fibre-optic cable is responsible for 99% of international telecommunications because it is still more reliable, more efficient, faster, and cheaper than satellite communication. In addition, the last two decades have seen the rapid expansion of submarine pipelines to keep pace with developments in the global oil and gas industry and its offshore rigs. More recently, we saw an explosion in the extensive cable networks that supply us with wind energy from offshore wind farms. Wind energy (and, to a lesser degree, wave energy) has plenty of potential to meet the growing global energy demand while reducing carbon emissions. Most turbines and large-scale wind farms are currently located near our coasts, but recent research shows that wind and even solar energy can be farming on the open sea, with mechanisms such as floating installations.

Other sectors also demand their place in or at the sea. Sea and coastal tourism is the second-largest employer in the ocean economy and one of the fastest-growing segments in the global tourism economy. More than 40% of the world's population lives within 100 kilometres of the ocean, and 12 of the world's 15 megacities are situated on the coast. As populations, economic activity, and urbanisation continue to grow in coastal areas, land reclamation – spraying sand to raise the surface of the land – has become critical in creating additional space. There are many famous examples of relatively small-scale land reclamation projects (harbour expansions and palm tree-shaped islands), but the world leader in large-scale reclamation projects is China. Its coastline grows by several hundreds of square kilometres every year.

Fortunately, there is also a growing group of people who recognise the dangers of this sometimes unbridled mass exploitation of resources and want to limit its extent. One way of doing so is to designate certain marine areas as protected areas (see below). This increase in protected areas is certainly a positive development, although in practice the ocean space remains geopolitically highly sensitive and, therefore, subject to change. Large parts of the ocean are divided into different maritime jurisdictions, allowing various countries to exert influence in conducting their economic and military activities.

Over which parts of the sea do countries have legal jurisdiction? The first 12 nautical miles (about 22 kilometres) off the baseline or low-water line along the coast are designated as 'territorial waters'. Generally, a coastal state has the same legal rights in its territorial waters as it does on land, with its own laws and administration of justice. In theory, vessels from other countries also have free passage through these waters. From there until a maximum of 200 nautical miles or 370 kilometres off the coast lies the country's exclusive economic zone (EEZ), within which a country has limited rights (such as the exploitation of resources, fishing, and scientific research) and duties (such as nature conservation) as established in the UN Convention on the Law of the Sea. Anything beyond this limit is referred to as 'international waters' or the 'open sea'. This covers 213 million square kilometres, 59% of the ocean's surface. And here, the general rule of *mare liberum* applies. In other words, the sea is free for seafaring and trade, a legal principle that dates to the writings of 17th-century Dutch scholar Hugo Grotius.

In practice, geopolitical discussions abound as to the exact boundaries and jurisdictions, which are not as clearly defined as they are on land. The South China Sea is a regular source of rising regional tensions sparked by its rich natural resources and fishing grounds, which are hotly contested by China, Malaysia, Brunei, the Philippines, Taiwan, and Vietnam. This region is so critical to fishing and shipping that an escalating conflict could have significant consequences, even for the world economy. Similar conflicts could arise in the Arctic, where the melting ice caps are revealing new shipping routes to areas with large oil and natural-gas reserves.

The recent discovery of precious metals and other resources on the seabed has led many countries to claim extensive rights over ocean space, with the inevitable discussions that arise as a result. Manganese nodules, for instance, are usually found in international waters, and so the International Seabed Authority (ISA) was established to determine who would be allowed to exploit the seabed and where. No satisfactory solution has yet been found. More problematic than the question of who gets to exert their rights on the deep-sea floor is the question of how much damage such activities would cause in the deep sea and what the consequences would be for the global ocean and its ecosystems.

GOLD RUSH AT THE BOTTOM OF THE SEA: DEEP-SEA MINING

If we look at things rationally, it makes perfect sense: as we become more dependent on electronics and need chips and batteries to power our electric cars and solar panels, we need to get our resources from *somewhere*. Those are primarily metals that are traditionally mined from the land. But as it turns out, the bottom of the sea also has countless manganese nodules, 'underwater crusts', and hydrothermal vents that contain valuable metals. Why should we not exploit those as well?

Since the 1960s, manganese nodules (also called polymetallic nodules) have been a strong focus of interest for governments and businesses looking for alternative ways to extract resources. The fact that they contain not only manganese but also nickel, copper, cobalt, and other precious metals as well as rare-earth elements makes them particularly interesting. However, initial attempts to turn this into a commercial success story failed. It was simply too difficult to harvest the manganese nodules in an economically viable way. This is now starting to change. Various trials have shown that robots can harvest the nodules using suction hoses. Several countries are currently engaged in a race to obtain these deep-sea riches in a commercially profitable manner. It is only a matter of time before these riches are dredged up to the surface. And the logical question that follows is: do we want that to happen? The answer is probably equally unsurprising: as tempting as it is, it is probably not a good idea to disrupt the complex world of the deep-sea floor. There are at least two reasons for not doing so: because of what we do know about life on the seabed and – just as critical – because of what we do not yet know about life on the seabed.

You may remember how endlessly deep and dark the deep sea is. Pressure there is so great that a human being would be crushed in an instant. The bottom does have some oxygen that is carried there via the deep-sea currents through the thermohaline circulation, but the layers above the sea bed have little to no oxygen due to the lack of sunlight for photosynthesis. That organisms can survive in such challenging circumstances is, in scientific terms, enough reason not to wipe them out before we have had a chance to study them. Moreover, we do not know what impact deep-sea mining will have on local ecosystems. That is why we believe we must adopt the precautionary principle: as long as we cannot prove that deep-sea mining will not catastrophically affect marine life, it is better not to do it. The burden of proof – extensive research into the risks involved – lies with the companies that want to disrupt a complex, extremely old ecosystem that encompasses a vast area.

The manganese nodule fields form an ancient, pristine ecosystem that has remained undisturbed by human activity up until now. It is estimated that only a tenth of all marine life around these manganese nodules has been identified. Meanwhile, we also know how complex ecosystems can hold surprising benefits for us, as was the case with sea-based medicines, phytoplankton, and the biological pump.

Some things we do know. The areas where the manganese nodules are found are teeming with life. The Clarion Clipperton Zone alone, an area in the north-eastern part of the Pacific Ocean that has been designated as the first area for commercial deep-sea mining, contains tens of billions of tonnes of manganese nodules spread over an area spanning more than four million square kilometres. Countless fascinating creatures live there: sea cucumbers, brittle stars, sea anemones, sponges... The nodules form a hard substrate in the otherwise muddy plains, to which marine animals can attach. Many organisms could suffer detrimental consequences from the sediment that is stirred up during harvesting, the silt that is spouted back into the water column, and the accompanying waste, noise, and light. Furthermore, these manganese nodules are not a renewable resource: once they are harvested, they are gone forever – in this respect, deep-sea mining is no different to mining on land. The ecosystem will not recover. A recent study hypothesises that the polymetallic nodules are a source of 'dark oxygen' (oxygen produced without light), which provides the sea floor in the deep ocean with oxygen. The process is largely unknown, and more research is being organised today, but the idea that oxygen is produced on our planet non-biologically is a game changer and a strong argument for a moratorium on deep-sea mining. In addition, we do not think it is wise to stir up sediment and the carbon that has been stored in sediment layers on the sea bed for millions of years. This too could have unforeseen consequences in distant areas, even up into the twilight zone. As far as we are concerned, all this is reason enough to refrain from destroying a habitat that is millions of years old by extracting manganese nodules.

The same applies to the minerals found around hydrothermal vents: now that we have caught a glimpse into the cradle of life, it would be irresponsible to exploit these intriguing ecosystems and do the same thing we did on dry land: cut down prehistoric forests and exterminate countless species. We should not make that mistake again, even if it is for a good cause such as a transition to renewable energy sources. Of course, mining on land also causes extensive damage, making it tempting to find an alternative in the deep sea. However, as long as no extensive research has been conducted into the long-term effects of mining operations in the deep sea, we strongly recommend refraining from inflicting damage there. We will have to look for better alternatives. We should theoretically have enough resources on land if we can make the transition to a circular economy with recycling and urban mining, which involves extracting resources from electronic waste. We can also develop technology that requires fewer precious resources, like cobalt-free batteries for electric vehicles.

THE BLUE ECONOMY: DIFFERENT INTERPRETATIONS AND INTERESTS

The example of deep-sea mining makes it clear that the blue acceleration does not automatically provide benefits. That also applies to related concepts such as 'blue growth' and the 'blue economy'. The common denominator underpinning these terms is the new role that the ocean is expected to play as the engine driving human development. It sometimes seems like the future of humankind does lie in the sea, which is considered an area for exploitation. The challenge will lie in ensuring that the ocean remains healthy and that its riches are used sustainably and equitably.

Lack of knowledge is not the only problem. There are also contradictory interests and visions, all with valid arguments that sometimes leave us with a difficult dilemma in choosing between economic interests and sustainability. Even within the movement towards greater sustainability, fierce discussions take place, as with deep-sea mining: what may initially look like the solution for a sustainable transition towards an economy free of fossil-fuel carbon and with electric vehicles may have disastrous consequences for marine life.

Whatever the case, we must be more transparent about the money flows involved with the main actors in the blue acceleration and impose stricter criteria to ensure that the revenue is spent sustainably. We also should not lose sight of social justice. The main beneficiaries of the blue acceleration are primarily economically powerful states and companies, while its detrimental effects will mostly be felt by developing countries and local communities. The vulnerability of the smaller island states and the least developed countries to the impacts of climate change will only increase with the blue acceleration. A growing number of studies emphasise that social and justice issues are just as important as environmental considerations in discussions about the future of our ocean.

Although we avoid getting involved in political discussions, it is our duty as scientists to provide information that is as accurate as possible. It is the only way to have a public debate based on facts. But even with incomplete facts we need to make judgment calls, simply because doing nothing is also a choice, immersed as we are in the blue acceleration.

The direction we should be heading seems clear: more sustainable forms of energy and industry so that we can reduce emissions of CO_2 and other harmful substances on the one hand, and promote healthier ecosystems on the other. But the road to get there is riddled with obstacles and scientific or ideological discussions. The debate surrounding wind farms has proven that technology is a double-edged sword with both benefits and drawbacks. Opponents are afraid that we are turning the North Sea into an industrial park and focus on its drawbacks: they are a form of visual pollution and endanger marine life. There are also concerns about the harmful effects on seabirds and bats, for instance. Vibrations caused by the construction and operation of wind farms can be hazardous for certain species, and some studies indicate changes in currents and sediments that could be harmful to phytoplankton. But these are preliminary studies, and there is no certainty yet. On the other hand, wind farms also provide positive benefits for biodiversity because they are undisrupted zones where fishing is not carried out in practice. So the positive aspects can also benefit other parts of the sea. In addition, wind farms are helpful in the transition towards more sustainable energy sources and fewer carbon emissions, which also indirectly benefits marine life.

Various components of the 'blue economy', such as fishing, aquaculture, transportation, tourism, and land reclamation, have more detrimental effects than wind farms do. As we said earlier, they take up increasing amounts of space. Our hope that the ocean will provide for our needs and those of future generations also entails potentially great risks. To be clear: the terms 'blue economy' and 'blue growth' are interpreted differently by different people and used by different organisations to advance different interests. Where one person may interpret it as all economic activities pertaining to the ocean, another may only include sustainable economic activities. And even 'sustainable' does not always mean the same to everyone.

Meanwhile, the world's seas are increasingly becoming a scene of quite a lot of economic activity. The 2020 EU Blue Economy Report estimated that the blue economy in Europe accounts for around 750 billion euros in revenue and 5 million jobs. Those numbers will have increased since then. Europe is also the world leader in technologies for harnessing the ocean's energy and hopes to draw 35% of its electricity from the sea by 2050. The European Green Deal underscores the continent's ambitions for sustainable growth as it employs the opportunities presented by coastal and offshore areas. In this case, their goal is not only a green but also a blue transition that creates new job opportunities and moves away from past mistakes – blind exploitation without taking into account the environment and social well-being in general. Ideally it would be a flourishing economy in a healthy context that respects both humans and the environment.

LOOKING TO THE FUTURE

The blue acceleration steams ahead mercilessly, with all its benefits and drawbacks. Pressure on the ocean is building on several fronts while the sea itself is changing. The number one cause of this shift is climate change, but also fishing, the expansion and urbanisation of coastal areas, the space taken up by the blue economy, international trade, and various forms of pollution.

These massive changes in the ocean are potentially unpredictable, capricious, and irreversible. Some of them we understand well, such as rising surface-water temperatures. However, most of them – acidification, the effects of microplastics and various chemicals, the processes occurring on the seabed, biodiversity erosion, and changes in ecosystems – have not yet been sufficiently studied. Some of these mechanisms are known, but their effects, such as rising sea levels, are hard to predict.

Marine science will need to step up the pace if it wants to keep up and keep in step with the blue acceleration. Although we know enough to warn of the consequences of various developments that worry us, there are still many gaps in our knowledge of the ocean that need to be filled. To start with, we need to better monitor our ocean and the marine life within it, both in terms of space and over time. A digital representation will allow us to better understand, manage, and predict how the sea works. Systematic, sustainable observation systems, with on-site sensors that can measure and quickly transmit reliable data from our ocean, seas, and shores, form an essential part of the solution. A number of new technological developments can be used to this end.

For instance, we are expecting a lot from the Argo fleet, which currently consists of 4,000 floats providing a constant stream of live data from different locations. They track a variety of metrics such as temperature, salt content, currents, and so on in the upper 2 kilometres of the ocean. They will soon also be able to chart the effects of rising sea temperatures on microbes, the biological pump, and oxygen loss, as well as monitor sunlight, oxygen levels, dissolved particles, chlorophyll, and nitrogen compounds. Just like the current floats, vast numbers of these new ones will be deployed from ships, after which they will be carried along by deep-sea currents at a depth of 1,000 metres. Every two days, they will use an oil-filled bladder to sink to a depth of 2,000 metres before rising back to the surface to transmit their data. In a more distant future, we hope to create a fleet of small robots that can travel even deeper and others that can take automated measurements in shallower coastal areas.

This fascinating data can be supplemented with other valuable technology, such as sonar, which will allow us to chart the seabed and its habitats as well as view schools of fish or plankton blooms within the water column. Other observational techniques include video imagery and acoustic cameras, hydrophones for measuring underwater noise, and even molecular techniques for real-time measuring of biodiversity. In addition to these new technologies, observations from traditional research ships will continue to be important.

In addition to the floats, other robots, such as small remotely operated submersibles, will also play a key role. Remarkably, we can draw inspiration for robot designs from the sea itself. The same goes for other materials that we call 'biomaterials' or 'bio-inspired design'. Marine organisms live in challenging, often extreme conditions in terms of salt content, pressure, light, and temperature, which has led to an immense diversity in building plans, physiological solutions, and behavioural patterns. Marine life has already inspired innovations in attachment mechanisms, anti-fouling coatings, armour, flotation capability, mobility, sensors, and invisibility.

Of course, all the observations and data collected about the sea have to be processed, which is no mean feat. One source of help in processing all this big data is artificial intelligence, a technology that is revolutionising how we process data. A number of scientists are currently working on creating a 'digital twin' of the Earth, a model of our planet that is as accurate as possible. A digital twin of the ocean is a realistic goal that is also being worked on in Europe. This twin is designed to simulate the ocean and its interactions with the atmosphere, the deep earth, ice, and land in detail. The data for this model comes from a variety of disciplines: physics, chemistry, biology, geology, bathymetry, satellite images, and the habitats themselves, but human activity at sea is also being mapped. The twin should ultimately allow us to make predictions regarding weather and natural phenomena; prevent algae blooms from forming; increase the sea's productivity; and make fisheries management, shipping, and marine operations safer.

This is all well and good, but to achieve this, we will need to move beyond traditional academic disciplines and work as a multi-disciplinary whole. At the Flanders Marine Institute (VLIZ) in Belgium, we are actively trying to 'marinate' other research disciplines to convince them to venture into the marine-research field.

In the medium term, three research domains will become critical: the ocean and climate change, sustainable living resources (particularly fisheries and aquaculture), and the impact of human activity on the ocean. The new technologies that we have described above will help us secure the future and health of the ocean, but we need to work together very closely to make it happen.

Whatever course we decide upon, the future holds fascinating times ahead for marine science.

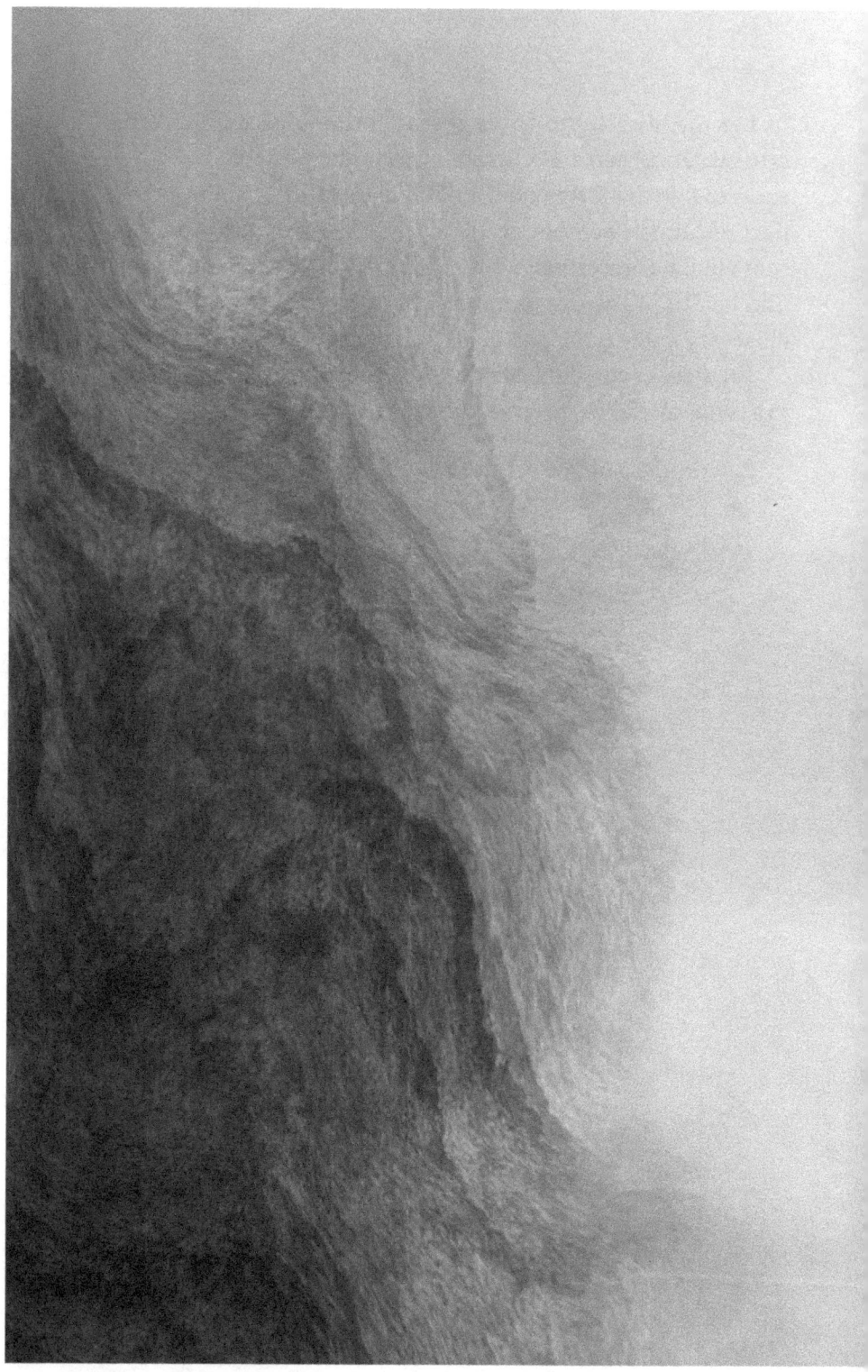

EPILOGUE

The ocean is what makes our planet unique in the solar system. The Earth is a water world. Our closest sister planets, Venus and Mars, once held water, but they dried up into the arid celestial bodies we know today. If the ocean is crucial for life, then life may not be limited to our planet with its specific ocean. Under the right circumstances, such as hydrothermal vents filled with metals and other elements, there is a chance that life emerged at some point in the past or may do so in the future. For us Earth dwellers, though, this is not much of a consolation. Even if there are planets out there with the right atmosphere and temperature, they are probably too far away for us to reach.

Therefore, we have no other choice but to cherish our planet, that vulnerable pale blue dot in space with its ocean, an ocean that holds the largest variety of ecosystems, produces half our oxygen, stores a quarter of the carbon humans produce, regulates the global climate, and preserves life on Earth. Without the ocean, the Earth would be an uninhabitable place. The sea has so much more to offer than fishing, shipping, and seaside holidays. And yet, little of that immense importance is reflected in our educational system or the general media.

TURNING THE TIDE

With this book, we hope to have made it clear that the ocean holds fascinating beauty which is worth protecting, but that it is currently under threat. Because of the effects of climate change, the cold part of the Gulf Stream is floundering, extreme events such as storm surges and (marine) heat waves are intensifying, sea life is increasingly under pressure due to oxygen loss and acidification, coastal defences and food supplies from fishing and aquaculture are threatened, and the biological pump is sputtering. In addition, pollution and overfishing are piling even more pressure on marine life. This combination of threats is a perfect storm that leaves us reeling, and no one can say with certainty that we will emerge unscathed.

In 2022, the United Nations declared an ocean crisis: we must turn the tide. Rising sea levels, rising ocean temperatures, and greenhouse-gas concentrations are breaking records every year. Lower-lying countries and coastal cities are faced with flooding, pollution and warming leading to 'dead zones', and overfishing is not under control. Marine pollution is increasing as marine species decline. Shark and ray populations have decreased by 70% in the past 50 years. Almost 80% of global untreated wastewater ends up in the sea, and every year, 8 to 10 million tonnes of plastic are added to the soup.

Most of the ocean is feeling the effects of human influence. In coastal areas, no less than three-quarters of all marine life is suffering from habitat loss. The global loss in coral and kelp (large algae) is estimated at 50 and 40%, respectively. More than 25 marine species have become extinct, and this number is expected to rise rapidly in the near future. Twenty to 25% of marine species are at risk of extinction, including corals, fish, seagrasses, and mangrove forests. In addition to the loss of biodiversity and habitat, the resilience of species and ecosystems is also decreasing, making them more vulnerable to disruptive events.

These problems are the result of humanity's colonisation of nature. Agriculture and the Industrial Revolution have accelerated this evolution. Humans dominate, change, and mould nature to their own needs. In part due to the rapid growth in populations and consumption, this has led to the destruction of nature at an unprecedented scale. These problems may initially seem to loom impossibly large over our heads like a giant tidal wave threatening to wash us away. But, at the same time, the reality is both more frightening and more comforting: this evolution is already fully underway, but it is a gradual process that can be guided by humans.

The good news is that people are listening to these warning signals. There has been a lot of talk about climate agreements, the protection of marine zones in areas beyond national jurisdiction, the banning of detrimental fishing subsidies, and the protection of biodiversity by designating protected areas. But ambitions are low, and there is little concrete action in practice. We need to raise the tempo a couple of notches. One key mindset change is not to see the sea simply as something to exploit.

To ensure that our planet remains a safe haven for humans, we must first manage the ocean as a common good for humanity. But how can we do so in concrete terms? As you may have guessed by now, the solution is not straightforward. Several years ago, the Ghent University administration building was marked with the following quote: 'The next big thing will be a lot of small things.' That eloquently sums up how we feel about the future of our ocean. There is no one-size-fits-all solution to all our problems. It would be poor risk management to focus all our efforts on one type of technology or approach because the chance of failure would be too great. Moreover, the possibility that taking one type of action would be enough is very slim. Technological solutions require an accompanying change in behaviour if we want them to be effective, just like behavioural change needs technological innovation to have an impact. The problems are simply far too big and complex.

SOLUTIONS

What can and should we do? First of all, we need to reduce human impact, and protect and restore our marine ecosystems. If we want to achieve a balance between people (social equality), planet (ecological health), and profit (economic wealth), we urgently need to restore balance to the environmental side of the equation. Just like with the current global climate crisis, societal awareness, political will, and the mobilisation of means will be necessary. In any case, 'mitigating' measures are required for the general climate problem: the energy transition to fewer fossil fuels, the food transition to less cattle farming and better land use, and the industrial transition to fewer carbon emissions. But we should also stop harmful activities at sea. Below are listed several measures that we feel are essential.

Reduce the pressure on fishing
The fishing industry is the world's largest industrial marine-based employer: it supplies millions of jobs and is an essential food source for billions of people. But, as we have seen earlier,

it is also by far the biggest threat to marine ecosystems. This may seem self-evident, but we must reduce our catch if we want to maintain healthy fish populations. We could do this by monitoring species extensively, but it also involves ceasing all non-sustainable fishing activity, including industrial fishing practices such as bottom trawling, large fishing fleets and catching fish in vulnerable zones or seasons, illegal fishing and harmful subsidies, and violation of fishing agreements. We can combat the latter by implementing better monitoring technology. The sea is currently the Wild West: less than 1% of its waters are protected, and more than half the fish stocks exploited by humans are overfished or exhausted.

Protect marine areas
There are currently 18,000 marine reserves, but many of them are ineffective due to a lack of management, lack of enforcement, or poor design. Fishing is even allowed in many of these so-called marine reserves. Most of them are 'paper parks'. The UN treaty that aims to protect 30% of the open seas by 2030 is a step in the right direction, but that does not apply to the often vulnerable coastal zones, which fall within countries' exclusive economic zones. In other words, the treaty needs to take into account the fact that the ocean is not a uniform body of water but that it contains ecosystems that are not spread equally over the seabed and its massive volume. The treaty must also be observed in practice. In other words, it should 'have some teeth'.

An earlier significant step in the protection of marine environments was the 1982 United Nations Convention on the Law of the Sea, which did not come into force until late 1994. This treaty deals with issues such as the determination of boundaries and the establishment of exclusive economic zones for each country, as well as scientific research and the protection of marine environments. But in practice, it has proved insufficient to protect aquatic life. More concrete action was and still is needed, including habitat restoration.

Restore damaged habitats
The best way to restore damaged habitats is to simply leave them alone. We can achieve this through a global ban on bottom trawling and other harmful fishing practices, a moratorium on deep-sea mining, less pollution, cleaner shipping, and a limit on coastal development. However, in addition to passive restoration, active restoration is also an option. These are often ambitious projects, such as building natural coastal defences such as dunes. Active restoration can also mean the re-introduction of species by planting mangrove forests and seagrass meadows, moving coral reefs, or building oyster banks. Removing pollution and limiting invasive species can also help. Another, more controversial, option is to use genetic modification to cultivate more heat-resistant coral, for instance. But remember: the cheapest option by far is to prevent damage from occurring in the first place.

Mitigate climate impact
Unfortunately, we cannot stop climate change, but we can mitigate its impact – and that includes its effect on the ocean. We can do so by restoring habitats as described earlier and by maintaining healthy marine ecosystems. Examples include seagrass meadows, mangrove forests, coral reefs, shellfish banks, salt marshes, and kelp forests, but also phytoplankton, krill, whales, and the deep sea: we now know that these are all essential carbon reservoirs.

Given the vast extent of climate change, it is also tempting to look to technological solutions. However, many of the potential solutions proposed by scientists or policymakers have drawbacks or potentially damaging results that need to be considered as well. We have already seen several initially promising examples, such as geo-engineering through the fertilisation of phytoplankton with iron or introducing olivine for carbon storage. Their feasibility on a large scale is uncertain, while the consequences, the potentially undesirable side effects, are impossible to predict. Moreover, the impact of climate change is so great that this alone is not enough. We need to cut back on emissions first.

We have determined that technology alone is not enough; it must be linked to concrete action. That task falls to governments and policymakers, but *we* can also contribute. How? By using water responsibly, adopting a climate-friendly lifestyle, advocating for the maintenance and recovery of coastal ecosystems, choosing sustainable seafood, and avoiding unnecessary waste… These are just a few of the steps you can take for our ocean and our planet. And – just as important – remember to every once in a while take the time to enjoy the majestic marine landscape and to show respect for what the ocean does for us without asking anything in return.

BLUE HOPE

Thankfully, we have made some great strides in environmental awareness and in making people realise the value of marine life. In that respect, it is worthwhile to watch The Silent World, the award-winning 1956 documentary by Jacques-Yves Cousteau and Louis Malle, the second film ever to show ocean life in all its colourful splendour. We now know that the ocean is anything but a silent world, but there is another reason why this documentary is considered outdated today. In one scene, employees blow up a coral reef with dynamite so they can count and study the fish. In another scene, their boat, the Calypso, hits a young sperm whale and fatally wounds the animal. The crew shoots the whale to put it out of its misery. When sharks arrive to feed on the carcass, Cousteau keeps the cameras rolling while the crew slaughters the sharks with harpoons for no apparent reason. These images are painful to watch today. Still, we should place the documentary in the context in which it was filmed: The Silent World dates from 1956. Jacques-Yves Cousteau did get involved in the protection of the environment, and times would also fundamentally change later.

There is also more awareness today where waste is concerned. Less than a century ago, governments still thought that dumping 30,000 tonnes of toxic war ammunition off the coast of Heist in Belgium was a good idea. In the 1980s, ships would leave port carrying titanium dioxide (a substance found in paint) or light radioactive waste for 'controlled' dumping several dozens of kilometres off our coast and in the Gulf of Biscay, respectively. We know better now, although new sources of pollution have since reared their ugly heads, with plastic waste on land, in the air, and in the sea being the biggest problem.

Marine science is an indispensable ally in our care for the planet, as this book has hopefully shown. We know only a fraction about the way ocean ecosystems function. Still, we know enough to see how crucial they are in regulating climate change and maintaining life. We now also know enough about what we do not know to move our research in the right direction. In the decades to come, we expect great things from cutting-edge technologies such as submersible robots, floats and buoys, sensors, and observation systems to monitor the ocean and find renewable energy sources from the sea. Artificial intelligence will also help improve and support existing technologies in processing the immense amounts of data involved. We will finally be able to chart the deep sea and the pelagic twilight zone, an essential water layer at a depth between 200 and 1000 metres about which we currently know very little. Fascinating times await us.

'There are many strange and wonderful things, but nothing more strangely wonderful than man,' the chorus sings in Sophocles' ancient Greek tragedy *Antigone*. The ambiguous 'strange and wonderful' is reflected in our nature: humans are capable of acting 'strange' and inflicting violence on themselves and the world around them, but they can also achieve 'wonderful' things. The fact that scientific and technological evolution is possible, as we have outlined above, shows that we can change course and that there is room for hope – provided we act in time.

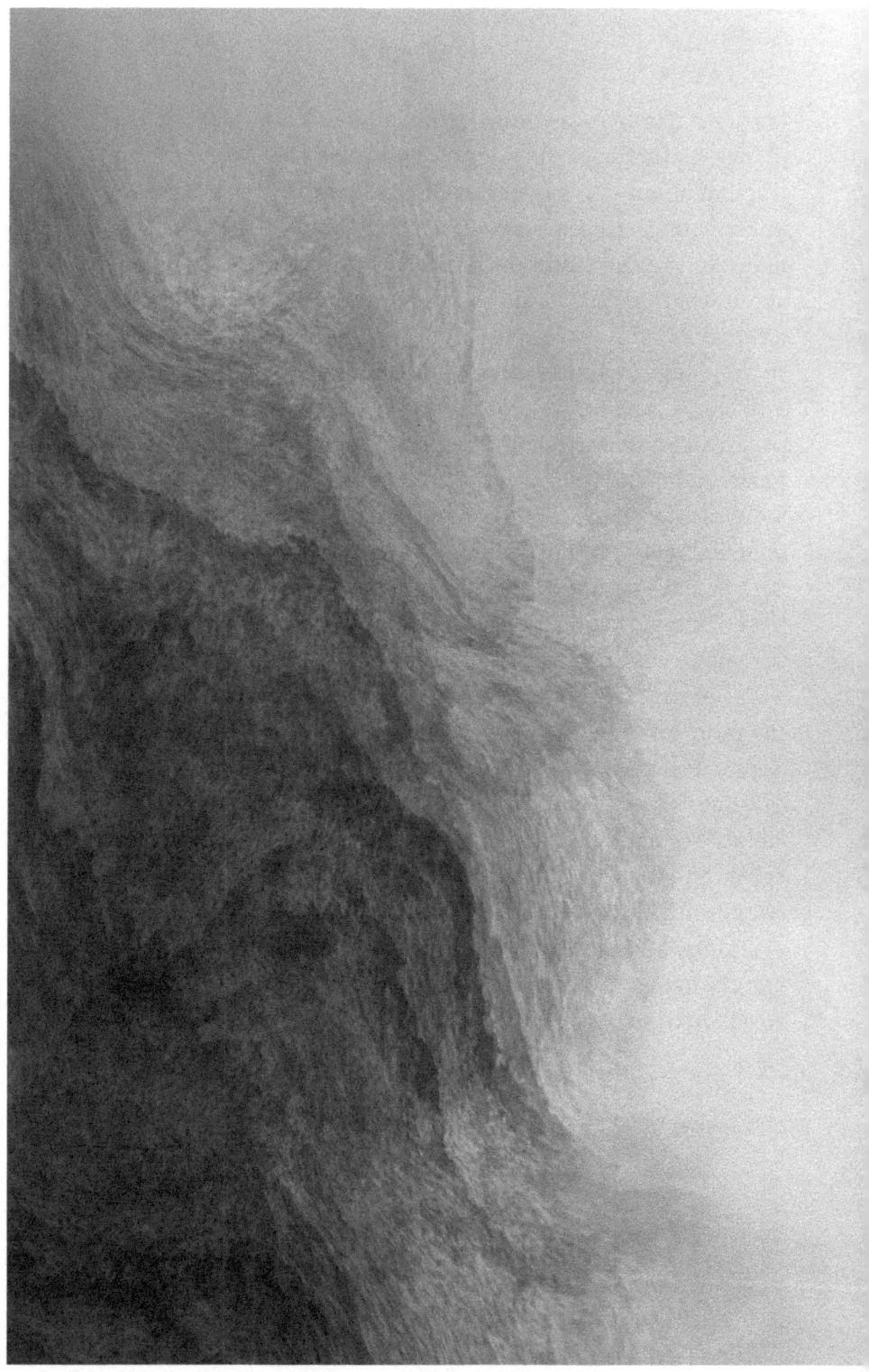

ACKNOWLEDGE-
MENTS

> "Homme libre, toujours tu chériras la mer."
> **CHARLES BAUDELAIRE**

The following scientists have read chapters from this book and offered valuable comments on specific chapters and paragraphs: Jana Asselman, Johan Braeckman, Peter Landschützer, Ann-Katrien Lescrauwaet, Florias Mees, Tine Missiaen, and Ilias Semmouri.

We want to thank our institutions, the Flanders Marine Institute (VLIZ) and Ghent University, in particular the Faculties of Sciences and Bioscience Engineering, for their support, their inspiration, and for giving us the freedom to write this book.

Britt Lonneville, VLIZ's GIS expert, created the maps for the Spilhaus Projection, the hydrothermal vents, the Challenger expedition, and the oceanic zones. Silke Lambert, PhD student at the Ocean & Human Health department at Ghent University, created the infographic about sea-spray aerosols. Chilekwa Chisala and Fons Verheyde from the VLIZ Marine Library tirelessly and quickly provided us with our requested articles and books.

We would also like to thank Isaac Demey from Academia Press for the excellent collaboration, his language skills, and his immense help in making our texts readable, as well as Meskerem Mees for her drawings marking the beginning of each chapter.

Marleen De Wit and Anne-Marie Ketels, thank you for your encouragement, patience, and feedback.

BIBLIOGRAPHY

Aldersey-Williams, H. (2016). *Tide. The Science and Lore of the Greatest Force on Earth*. W.W. Norton & Company: New York.

Amkreutz, L. & Van der Vaart-Verschoof, S. (2022). *Doggerland: Lost world under the North Sea*. Sidestone Press: Leiden.

Amon, D.J. et al. (2022). Assessment of scientific gaps related to the effective environmental management of deep-seabed mining. *Marine Policy* 138: 105066.

Appeltans, W. et al. (2012). The magnitude of global marine species diversity. *Current Biology* 22: 2189-2202.

Asselman, J. et al. (2019). Marine biogenics in sea spray aerosols interact with the mTOR signaling pathway. *NPG Scientific Reports* 9(1): 675.

Austen M.C. et al. (2019). *Valuing Marine Ecosystems – Taking into account the value of ecosystem benefits in the Blue Economy*. Future Science Brief 5 of the European Marine Board, Ostend, Belgium.

Bagusche, F. (2020). *Het geheime leven van de oceaan*. Luitingh-Sijthoff: Amsterdam.

Barbier, E.B. (2017). Marine ecosystem services. *Current Biology* 27: R431-R510.

Bates, A.E. et al. (2014). Defining and observing stages of climate- mediated range shifts in marine systems. *Global Environmental Change* 26: 27-38.

Braeckman, J. & Van Speybrouck, L. (Ed.) (2022). *Fascinerend leven. Een geschiedenis van de biologie.* Academia Press: Ghent.

Branscomb, E. & M.J. Russell (2018). Why the submarine alkaline vent is the most reasonable explanation for the emergence of life. *BioEssays 2018*, 1800208.

Breyne, M. et al. (2010). The world's very first marine research station in Ostend (Belgium). *Earthzine August 18*: 4.

Camprubi, E. et al. (2019). The Emergence of Life. *Space Science Reviews 215*, 56.

Copernicus Marine Service (2022). *Copernicus Ocean State Report*, issue 6.

Costanza, R. (1999). The ecological, economic, and social importance of the oceans. *Ecological Economics 31*: 199-213.

Costello, M.J. (2015). Biodiversity: the known, unknown, and rates of extinction. *Current Biology 9*: R368-R371.

Devriese, L.I. & Janssen, C.R. (2023). *Overzicht van het onderzoekslandschap en de wetenschappelijke informatie inzake (marien) zwerfvuil en microplastics in België.* VLIZ Beleidsinformerende Nota's, 2023_002. Vlaams Instituut voor de Zee (VLIZ): Ostend.

Dauwe, S. et al. (2021). Mariene klimaatmitigatie: een wetenschappelijke synthese van de meest pertinente oplossingsrichtingen voor het Noordzeegebied. VLIZ Beleidsinformerende Nota's, 2021_003. Vlaams Instituut voor de Zee (VLIZ), Ostend.

Dauwe, S., Pirlet, H., Gkritzalis, T. & Landschützer, P. (2023). *The opportunities and challenges of marine carbon accounting - a case study for the North Sea shelf ecosystem and the potential value of the ICOS Oceans Network.* VLIZ Beleidsinformerende Nota's, 2023_01. Flanders Marine Institute (VLIZ), Ostend.

European Marine Board (2013). *Linking Oceans and Human Health: A Strategic Research Priority for Europe.* Position paper 19.

European Marine Board (2017). *Marine Biotechnology: Advancing Innovation in Europe's Bioeconomy.* Policy Brief No. 4.

European Marine Board (2019). *Navigating the Future V: Marine Science for a Sustainable Future.* Position Paper 24.

European Marine Board (2021). *Sustaining in situ Ocean Observations in the Age of the Digital Ocean.* Policy Brief No. 9.

Everaert, G. et al. (2018). Risk assessment of microplastics in the ocean: modelling approach and first conclusions. *Environmental Pollution* 242(B): 1930-1938.

Falkowski, P. G. (2015). *Life's Engines: How Microbes Made Earth Habitable.* Princeton University Press.

FAO (2022). *The State of World Fisheries and Aquaculture 2022. Towards Blue Transformation.* Rome, FAO.

Fockedey, N. et al. (2018). *Vis- en zeevruchtengids voor professionele gebruikers. Bewust kiezen voor duurzame producten uit de zee.* Ethic Ocean/VLIZ: Paris, Ostend.

Gattuso, J.-P. et al. (2018). Ocean solutions to address climate change and its effects on marine ecosystems. *Frontiers in Marine Science* 5, 337.

Georgian, S. et al. (2022). Scientists' warning of an imperiled ocean. *Biological Conservation* 272: 109595.

GESAMP (2019). *High Level Review of a wide range of proposed marine geoengineering techniques.* GESAMP Working Group 41.

Gibbons, A. (2010). The World's first fish supper. *Science*, June 1, 2010.

Global Carbon Budget (2022). *Earth System Science Data Discussions 2022*, 1-159.

Gruber, N. et al. (2023). Trends and variability in the ocean carbon sink. *Nature Reviews Earth & Environment* 4, 119–134.

IOC-UNESCO (2020). *Global Ocean Science Report 2020. Charting Capacity for Ocean Sustainability.* Paris, UNESCO Publishing.

IOC-UNESCO (2022). *State of the Ocean Report 2022. Pilot edition.* Paris, UNESCO Publishing.

IPCC (2019). *IPCC Special Report on the Ocean and Cryosphere in a Changing Climate.* Cambridge University Press: Cambridge & New York.

IPCC (2019). *Special Report on the Ocean and Cryosphere in a Changing Climate - Summary for Policymakers.*

IPCC (2021). *Climate Change 2021, the Physical Science Basis - Summary for Policymakers.*

Jouffray, J.-B. et al. (2020). The Blue Acceleration: The Trajectory of Human Expansion into the Ocean. *One Earth* 2(1): 43-54.

Kenny, A. (2006). *Ancient Philosophy. A New History of Western Philosophy, Volume 1.* Oxford University Press: Oxford.

Lane, N. (2016) *The Vital Question: Why Is Life the Way it Is?* Profile Books: London.

Lane, N., Allen, J.F. & Martin, W. (2010). How did LUCA make a living? Chemiosmosis in the origin of life. *BioEssays* 32: 271–280.

Leroi, A. (2014). *The Lagoon. How Aristotle Invented Science.* Bloomsbury Circus: London.

Lescrauwaet, A.-K. et al. (2010). Fishing in the past: Historical data on sea fisheries landings in Belgium. *Marine Policy* 34(6): 1279-1289.

Lescrauwaet A.-K. et al. (2013). Hundred and eighty years of fleet dynamics in the Belgian sea fisheries. *Reviews in Fish Biology and Fisheries* 23: 229-243.

McCauley, D.J. et al. (2015). Marine defaunation: animal loss in the global ocean. *Science* 347 (6219), 1255641.

Morris, R.L. et al. (2019). Design options, implementation issues and evaluating success of ecologically engineered shorelines. *Oceanography and Marine Biology: an Annual Review* 57: 169-228.

OECD (2016), *The Ocean Economy in 2030*, OECD Publishing, Paris.

OECD, 'Ocean Economy', 2024, https://www.oecd.org/en/topics/sub-issues/ocean/ocean-economy.html

Pauly, D. et al. (1998). Fishing down marine food webs. *Science* 279: 860-863.

Pauly, D. et al. (2002). Towards sustainability in world fisheries. *Nature* 418: 689-695.

Pauly, D. & Zeller, D. (2016). Catch reconstructions reveal that global marine fisheries catches are higher than reported and declining. *Nature Communications* 7(1), 10244 10244.

Pauly D. (2021). What Netflix's Seaspiracy gets wrong about fishing, explained by a marine biologist. *Vox*, April 13, 2021.

Pirlet, H. et al. (2022). *Marien onderzoek in Vlaanderen en België: Een inventaris van het onderzoekslandschap.* VLIZ Beleidsinformerende Nota's, 2022_004. Vlaams Instituut voor de Zee (VLIZ), Ostend.

Roberts, C.M. et al. (2017). Marine reserves can mitigate and promote adaptation to climate change. *Proc. Nat. Acad. Sci.* 114(24): 6167-6175

Rogers, A.D. et al. (2015). Delving Deeper: Critical challenges for 21st century deep-sea research. Larkin, K.E., Donaldson, K. & McDonough, N. (Eds.) *Position Paper 22 of the European Marine Board.*

Russell, M.J. (2018). Green rust: the simple organising 'seed' of all life? *Life 8,* 35.

Sala, E. et al. (2021). Protecting the global ocean for biodiversity, food and climate. *Nature* 592: 397-402.

Stewart, R.H. (2009). *Introduction to Physical Oceanography.* University Press of Florida: Gainesville.

Stow, D. (2014). *Oceans: A Very Short Introduction.* Oxford University Press: Oxford.

Sweetman, A.K. et al. (2024) Evidence of dark oxygen production at the abyssal seafloor. *Nat. Geosci.* 17: 737–739.

United Nations Office of Legal Affairs (2021). *The Second World Ocean Assessment.*

Van Cauwenberghe, L. et al. (2013). Microplastic pollution in deep-sea sediments. *Environmental Pollution* 182: 495-499.

Verleye, T.J. et al. (2022). *Het voorkomen van tsunami's, rogue waves en infragravitaire golven in de zuidelijke Noordzee - Een wetenschappelijke synthese.* VLIZ Beleidsinformerende Nota's, 2022_003. Vlaams Instituut voor de Zee (VLIZ), Ostend.

Von Schuckmann et al. (2020). Heat stored in the Earth System: where does the energy go? *Earth System Science Data* 12: 2013-2041.

Waterfield, R. (Ed.) (2009). *The First Philosophers: The Presocratics and Sophists.* Oxford University Press: Oxford.

Watson, R. & D. Pauly (2001). Systematic distortions in world fisheries catch trends. *Nature* 414: 534-536.

Weiss et al. (2016). The physiology and habitat of the last universal common ancestor. *Nature Microbiology* 1: 16116.

Wulf, A. (2015). *The Invention of Nature. The Adventures of Alexander von Humboldt, the Lost Hero of Science*. John Murray Publishers: London.

Zeller, D. & Pauly, D. (2019). Viewpoint: Back to the future for fisheries, where will we choose to go? *Global Sustainability* 2, e11, 1-8.

Zilhao, J. et al. (2020). Last Interglacial Iberian Neandertals as fisher-hunter-gatherers. *Science* 367: 1-15.

OTHER GENERAL RESOURCES USED:

Testerep magazine from VLIZ, Flanders Marine Institute: https://www.vliz.be/testerep (in Flemish).

De Grote Rede magazine from the VLIZ, Flanders Marine Institute: https://www.vliz.be/en/publications/type/grote_rede.

The Compendium for Coast & Sea: http://compendiumkustenzee.be/en and the Coastal Portal https://www.kustportaal.be/en.

VLIZ, Flanders Marine Institute historical figures: https://www.vliz.be/wetenschatten/scientists.php (in Flemish).

Academia Press
Coupure Rechts 88
9000 Gent
België

www.academiapress.be

Academia Press is a subsidiary of Lannoo Publishers.

ISBN 978 90 209 8519 1
D/2025/45/380
NUR 912, 930

Colin Janssen & Jan Mees
The unknown sea. The importance of the ocean for people and planet.
Gent, Academia Press, 2025

Cover and layout: Joost Van Lierop
Illustrations: Meskerem Mees
Maps and infographics: Britt Lonneville and Silke Lambert

© Colin Janssen & Jan Mees
© Lannoo Publishers

No part of this publication may be reproduced in print, by photocopy, microfilm or any other means, without the prior written permission of the publisher.

The Unknown Sea